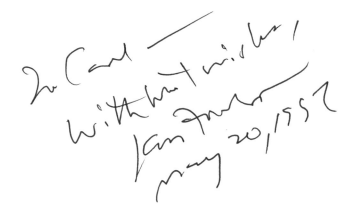

To Carl —

With best wishes!

Ian Fraser

May 20, 1997

THE FUSION QUEST

THE

FUSION

QUEST

T. Kenneth Fowler

THE JOHNS HOPKINS UNIVERSITY PRESS : BALTIMORE AND LONDON

© 1997 The Johns Hopkins University Press
All rights reserved. Published 1997
Printed in the United States of America on acid-free paper
06 05 04 03 02 01 00 99 98 97   5 4 3 2 1

The Johns Hopkins University Press
2715 North Charles Street
Baltimore, Maryland 21218-4319
The Johns Hopkins Press Ltd., London

Library of Congress Cataloging-in-Publication Data will be found
at the end of this book.
A catalog record for this book is available from the British Library.

ISBN 0-8018-5456-3

To Carol

Contents

Illustrations follow page 100

Preface & Acknowledgments

The Promethean quest to harness fusion energy, to capture the fire of the Sun on Earth, has sometimes been called the greatest technological challenge of all time. Now, after four decades of research, "controlled" fusion has at last been demonstrated in the laboratory. Having been privileged to participate personally in much of this exciting scientific history in the making, I was delighted when the Johns Hopkins University Press approached me about doing this book to share the fusion story with others.

I have three aims in writing this book. First and foremost, I am writing for anyone who wants to learn enough about the science of fusion energy to experience firsthand the exciting sense of discovery that comes from watching a new science unfold its mysteries. Most of the difficult, and therefore exciting, aspects of fusion research concern magnets and electricity, heat and light—phenomena that we all encounter in everyday experience and that make up the subject matter of classical physics. The difficulty lies not in the principles but in the complexity of execution. The task of this book, then, is to look beneath the complexity to help the reader see the fusion quest in familiar terms. Second, I am writing to young, aspiring scientists and engineers to say that it is not too late to be a part of the fusion adventure. Indeed, in light of what we know now, the best may be yet to come. And finally, mindful that four decades is a long time to wait, I am writing to the skeptical citizen or public official who, after all the time and money spent, wants to decide for himself or herself whether fusion will ever make it; and if it does, whether it will be all that its advocates have claimed.

This book is not intended as a textbook, though I have tried to retain scientific accuracy while explaining complex ideas in nonmathematical terms. In this, I am grateful to my students at Berkeley, who have insisted that what I say make sense, with or without mathematical language.

Nor is this intended as a book of advocacy for fusion, though I do hope that my own enthusiasm for fusion will be contagious. Since fusion is Big Science, fusion and governmental support are inextricably linked, now on a worldwide scale. Taking care to separate fact and opinion, I will touch upon these matters from time to time, drawing upon my nearly twenty years of experience as director of magnetic confinement fusion research at the Lawrence Livermore National Laboratory.

Fusion is a long-term proposition, a potentially abundant and environmentally attractive energy source to sustain industrial society in the twenty-first century and for millennia to come. The fuel supply is virtually infinite, and the radioactive hazards associated with fusion should be more benign and manageable than those of today's nuclear reactors, which utilize nuclear fission. Though fusion is the main energy source of hydrogen bombs, fusion alone has never produced a bomb; the hydrogen bomb requires a fission-based atomic bomb to set it off. But the virtues of fusion are offset by the fact that creating a fusion reactor is really difficult.

That nuclear fusion is the source of energy in the Sun and the stars has been known for more than sixty years. But would it work on Earth? Though this question was asked at the outset, research on fusion as a practical energy source only began in earnest in 1958, when research in magnetic confinement fusion, initially secret, was declassified worldwide. The new research program was called Controlled Thermonuclear Research, to indicate that it had no connection with hydrogen bombs.

Despite early optimism, by the mid-1960s the road to practical fusion power looked long and uncertain, if indeed there was a road. In 1966, some nine years after I entered the field, I co-authored an article in *Scientific American* on progress in fusion research. In these, our darkest days, my co-author and I asked ourselves if we could with confidence identify any design concept for a magnetic fusion reactor that was sure to work, even if large. Upon concluding that such a device might be the size of a small asteroid, we abandoned the idea and wrote instead about the scientific principles that gave us hope.

As it turned out, the principles we wrote about in 1966, together with experiments to test and quantify them, have indeed led researchers to the threshold of practical fusion power. The first successful attempts to produce controlled fusion energy on Earth occurred in brief experiments in the Joint European Torus (JET) in 1991 and then definitively at Princeton University in a series of experiments beginning in December 1993, and as this book is being written, work is under way to design the world's first experimental fusion reactor. Called the International Thermonuclear Experimental Reactor (ITER), this joint venture of the United States, the European Union, Japan, and Russia would, if actually constructed, be the largest international science project ever undertaken. In addition to further advancing the science of fusion, ITER will give researchers practical engineering experience with a fusion device approaching the size of an actual power plant. Meanwhile, great advances have been made on an entirely different approach, using lasers rather than magnets, and this also has a pending large facility, the National Ignition Facility (NIF).

The large size of ITER is necessary, ITER being our best answer today for a magnetic confinement design concept with a good chance of succeeding. What makes ITER so big? Will lasers or magnets get there first? Do the scientific principles of fusion hold promise for even better things to come?

To help answer these questions, this book will explore the principles upon which fusion is based, their experimental and theoretical foundation, and their implications for the design of fusion power plants. I will also discuss briefly fusion's potential to provide an environmentally acceptable new energy source for the twenty-first century, when society will be more vulnerable to energy shortages and energy-related pollution than most people realize.

We will learn about magnets nearly as cold as outer space, surrounding miniature "stars" hotter than the Sun; and lasers that for the merest split second produce a blinding flash of light more powerful than every light bulb in America turned on at once. We will trace the fascinating fabric of classical physics, from Newton to Einstein, from Faraday to Lorentz, as we piece together the principles of fusion science today.

The book is divided into four parts. Part I introduces the reader to fusion and the highlights of magnetic fusion research, culminating in the definitive demonstration of fusion energy at Princeton. Part II traces the development of the theory, technology, and engineering fundamental to magnetic confinement fusion, leading up to the Princeton event and preparing the way for ITER. Part III discusses the inertial confinement fusion approach and the NIF. Part IV places fusion research in its political context and concludes with a chapter on some possible technical breakthroughs on the horizon that could change everything.

Both the evolving science and the program history are presented from a personal viewpoint, as I experienced them, because this allows me to discuss what I know best. I hope that I have been able to share with readers most of the highlights of this wonderful adventure, and I extend sincere apologies to all my colleagues for the many omissions to which personal reminiscences fall prey.

I would like to thank many people who have contributed to this endeavor, starting with Valerie Wysinger-Razor, who typed the manuscript. I am grateful to Paul Rutherford and John Dawson for their review of and comments on the chapters dealing with magnetic fusion, and to John Holzrichter, Roy Johnson, Bill Kruer, John Lindl, and John Nuckolls for their patient tutelage in helping me through the section on inertial confinement fusion. Thanks go also to Stephen Craxton, John Holdren, Robert McCrory, Michael Perry, Tom Simonen, and Max Tabak for help with particular topics. Needless to say, all remaining errors are my own.

I am especially grateful to the graduate students for whom I have been thesis adviser or co-adviser: their work appears throughout these pages. Special thanks go to Joseph Fitzpatrick, Renato Gatto, Daniel Hua, Erfan Ibrahim, and Bijal Modi for their work in tokamak physics; to Cynthia Annese, Fred Brechtel, Zvi Covaliu, Jeff Latkowski, Ronnen Levinson, Micah Lowenthal, Quang Nguyen, Joel Rynes, Dennis Sarigiannis, Lawrence Sevigny, and Edward Watkins, for their work in fusion environment and safety; to Daniel Finkenthal, for his work in experimental diagnostics; to Nigel Barboza, for his work in inertial heavy ion drivers; to Scott Parker and Richard Procassini, for their work in numerical simulation; and to Eric Coomer, Karl D'Ambrosio, Gregory DiPeso, Jill Hardwick, Harry McLean, Uri Shumlak, and Russell Stachowski, for their work in spheromak theory and experiment. Thanks go also to my postdoctoral associates Alain Brizard, Shu Ho, and Taina Kurki-Suonio.

Finally, at the Johns Hopkins University Press, special thanks go to senior manuscript editor Miriam Kleiger for many improvements on every page of the text, and to science editor Robert Harington for encouraging me to undertake this book and for guiding it to its current form.

I

PROMETHEUS UNBOUND

1 : The Allure of Fusion

A ccording to Greek mythology, fusion energy, the fire of the Sun, is a gift hard won. When the Titan god Prometheus stole the Sun's fire to give helpless humans the power of survival, Zeus was so angered that he chained Prometheus to a mountain, where a vulture tore away at his liver for a thousand years. Today, irresistibly drawn to the challenge of bringing fusion energy down to Earth from the stars, scientists tempt Zeus still. Recently, after forty years of strenuous effort, fusion power was finally demonstrated in the laboratory, first in the Joint European Torus in 1991, and then definitively at Princeton University, in a series of experiments that began in December 1993. This book traces the scientific journey that led up to this exciting achievement, and examines what lies ahead and the impact that fusion could have on the survival and well-being of industrial society in the future. It is not an accident that, in Greek, the name Prometheus means "forethought."

Predating even Prometheus and the Sun, fusion is as old as the universe. Just a few minutes after the Big Bang, when time began, the unbelievably hot universe had expanded and cooled sufficiently to allow hydrogen nuclei to form deuterium, and deuterium to fuse into helium. Though most of the deuterium was consumed in this early, cataclysmic burst of fusion reactions, a trace survived. It is this trace of deuterium, together with small amounts created in later astronomical events such as supernovas, that is the main motivation for fusion energy research today. This deuterium, itself a form of hydrogen, is a small constituent of all water, about one part in five thousand. Yet, given the vast amount of water available, even this trace amount of deuterium represents an almost limitless supply of energy on Earth.

The first speculations about how to tap this vast store of energy were made a few years after Albert Einstein's brilliant deduction, in 1905, that mass can produce energy, according to his famous formula $E = mc^2$. Since the days of the ancient Greeks, philosophers and scientists have sought the nature of things in tiny building blocks, called atoms, of which all material objects are composed. These atoms were thought to be indestructible. Then, in 1911, Lord Rutherford postulated the nuclear atom, consisting of a heavy nucleus with a positive electric charge, surrounded by a cloud of negatively charged electrons. Electrons had been discovered earlier, by

J. J. Thomson, in 1897. The ultralight electrons, jumping from one atom to another, would turn out to explain all of chemistry, the roaring energy of steam engines, automobiles, and blast furnaces, and all that constitutes the industrial age as we have known it. But the tiny, heavy nucleus of atoms held a secret store of energy far greater than anything ever imagined before.

The first publication tying Rutherford's atom and Einstein's theory to an explanation of the stars appeared at the start of the Great Depression, in 1929. On the basis of the known masses of atoms, the paper's authors, R. d'E. Atkinson and F. G. Houtermans, predicted that the fusing together of the nuclei of very light atoms (such as hydrogen) would produce new atoms that would weigh less than the total weight of the original constituents. According to Einstein's formula, this loss of mass should produce an enormous amount of energy, enough to account for the light from stars.

Even in 1929, the possibility of using this new kind of energy on Earth did not escape notice. In a monograph on fusion that he edited in 1981, the physicist Edward Teller relates the story that, when a young Russian, George Gamow, reported on the paper by Atkinson and Houtermans at a meeting, a leading member of the Communist party offered the entire electrical output of the city of Leningrad for one hour each night if Gamow would undertake to produce fusion energy in the laboratory. According to Teller, "Gamow, a physicist of unusual taste and common sense, did not accept the offer." In America, Eastman Jacobs and Arthur Kantrowitz of the Langley Memorial Aeronautical Laboratory did make an aborted attempt to produce fusion in the laboratory in 1939, but the project was soon canceled by the laboratory director.

While nuclear fusion reactions such as those speculated about by Atkinson and Houtermans were soon discovered, and Hans Bethe went on to produce the quantitative theory of fusion energy in stars that would earn him a Nobel Prize, fusion soon took a back seat to the discovery, in 1939, of the nuclear fission reaction, which led to the development of the atomic bomb during World War II and the fission-based nuclear reactors of today. To date, both fusion and fission are seen as a source of nuclear energy to produce heat, which in turn produces steam to drive electric generators. However, the nuclear energy itself is produced in very different ways in fusion and fission reactions.

As its name implies, nuclear fission is the splitting in half of a heavy nucleus, such as that of uranium-235 (so named because it has 235 times the mass of a hydrogen nucleus). The two halves, called "fission fragments," form lighter elements that are usually radioactive. This is the nuclear waste from today's nuclear reactors, which must be safely protected from accidents during reactor operation and eventually disposed of in deep underground vaults away from public exposure. By contrast, as we have seen, nuclear fusion is the combining of two very light atomic nuclei, usually forms of hydrogen, to form a heavier element. Though fusion is not totally devoid of radioactivity, its main by-product is usually helium, the harmless gas used to fill balloons. The other main difference between fusion and fission is that fission can be made to occur relatively easily, by assembling a sufficient mass of uranium or plutonium in one place, whereas fusion only occurs at the enormous temperatures found at the center of stars, or in a nuclear bomb blast. Thus, while both fission and fusion have been used in nuclear weapons, a fission bomb is needed to detonate a fusion bomb. It is these differences that lead us to believe that fusion power, if it could be developed, would give fusion advantages in terms of environmental acceptance. But it is because creating fusion is such a demanding task that fusion is still in the research stage, while fission-based nuclear reactors already provide about 20 percent of the electricity generated in the United States and 70 percent of the electricity generated in France.

The first public evidence that anyone was thinking about fusion reactors appeared in the newspapers in March 1951, when I was a sophomore engineering student at Vanderbilt University. President Juan Perón of Argentina had claimed that a German scientist in his employ had produced fusion in the laboratory. While I do not remember seeing this in the Nashville newspapers, and the *New York Times* immediately claimed that such a thing would violate the laws of physics, an astute astrophysicist named Lyman Spitzer, tantalized by the claim, set about devising his own solution to the problem. Soon thereafter, with the approval of the U.S. Atomic Energy Commission, Spitzer established what would become the Princeton Plasma Physics Laboratory.

All of this was still unknown to me when I took my first job in 1957, fresh out of graduate school, in the still-secret fusion research program at the Oak Ridge National Laboratory. It is amazing now to think back to how little — and yet how much — we knew at that time. The seeds were there, many of them in a book Spitzer had published in 1956, which I devoured in every spare moment. Yet for every answer, ten new questions appeared. So that you can begin to share this experience, please join me now as we delve back in time to the discoveries that lay the groundwork for fusion science.

Let us start from the Rutherford atom, with its heavy nuclear core surrounded by electrons circulating around the nucleus, somewhat as planets circulate around the Sun. Picturing atoms in this concrete way is metaphorical, but useful. Looking more deeply, we find that the nucleus itself is a cluster of even smaller particles: protons, each with one positive charge (exactly equal to the negative charge of an electron); and neutrons, electrically neutral particles postulated by Rutherford but first discovered in the laboratory by James Chadwick in 1932.

The various chemical elements are then built up in the following way: First there is the nucleus of an ordinary hydrogen atom, consisting of a single proton and no neutrons. One proton with one electron circulating around it is a hydrogen atom. Adding one neutron to the hydrogen nucleus produces a deuterium nucleus. Because it still has only one positive charge, deuterium behaves chemically like hydrogen, and it is called an "isotope" of hydrogen. It is for this reason that deuterium is found thoroughly mixed with ordinary hydrogen in water; chemically, one isotope of hydrogen is as likely as another to join with oxygen to form water. Combining two neutrons with a proton gives yet another isotope of hydrogen, called tritium. To obtain heavier elements, we simply add more protons and neutrons, either one at a time, or by combining, or fusing, lighter elements.

As we shall see, the fusion process most likely to be used in the first fusion reactor is the combining of deuterium and tritium nuclei to make the common form of helium, helium-4. It works like this:

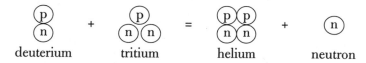

deuterium   tritium   helium   neutron

Here p stands for proton and n stands for neutron. Deuterium and tritium nuclei contain one proton apiece, and combine to form helium, which has two protons. The helium nucleus also contains two neutrons, one from the deuterium and one from the tritium; one neutron is left over, since tritium has two neutrons. We will call this the DT reaction.

This little sketch is the only nuclear physics we will need to know. I have used it many times to explain what fusion is all about to lay persons and students alike. This sketch can tell us about both the promise and the problems of fusion research. It simply says that fusing deuterium and tritium, which are forms of hydrogen, produces helium and a neutron.

The first point to be made from this sketch of the DT reaction is this: Though we have accounted for all the protons and neutrons making up the various nuclei, if we were to carefully weigh the deuterium and the tritium together (or more realistically, a large batch of deuterium and tritium in equal proportions), and then were to weigh the helium and the neutron, we would find that a tiny fraction of the original mass had been lost when the deuterium and tritium nuclei were fused to create the helium and the neutron; one helium nucleus plus one neutron weighs less than one deuterium nucleus plus one tritium nucleus. Of course, as we have seen, it is this loss of mass that accounts for the large energy release in the fusion process, according to Einstein. This energy shows up as the fast motion of the helium nucleus and the neutron, each moving at speeds exceeding 10 million (10^7) meters per second. (A meter is about 3.28 feet.) This corresponds to millions of times more energy than is produced by a typical chemical reaction such as the burning of gasoline; hence the enormous energy available in even the trace amounts of deuterium in water—just the tiny amount of deuterium in one gallon of water would produce as much energy as three hundred gallons of gasoline.

The second point is that the product of the DT reaction, helium, is neither radioactive nor chemically harmful. So the "ashes" from fusion are benign. Again, contrast this with the nuclear fission of a heavy nucleus such as uranium, which, when it splits, invariably produces smaller nuclei that are radioactive and therefore must be carefully disposed of as nuclear waste. Note, however, that DT fusion reactions do produce free neutrons moving at high speed. These fast neutrons create radioactivity when they bombard the materials—especially metals—of which the fusion reactor is constructed. Thus, while the fusion process does not produce nuclear waste directly, the fusion reactor itself does become radioactive, and its components must be disposed of safely when the reactor is finally shut down after the usual thirty- to fifty-year life of any electric power plant. In principle, this problem can be minimized by deliberately choosing construction materials that either produce less radioactivity or produce radioactivity that dies away more rapidly, so that the disposal deep underground that is required for fission waste is not needed. Choosing construction materials wisely is a vital part of fusion research. We shall return to this important point in discussing the environmental characteristics of fusion energy in chapter 15.

The last point to be made in this discussion of the DT fusion reaction concerns the positive charges of the deuterium and tritium nuclei that we would like to fuse together. Ordinarily, atomic nuclei, protected by their clouds of electrons, never come close enough to fuse, so fusion could never happen between ordinary atoms. Even if we strip away the electrons (not too difficult a task), the nuclei would strongly repel each other because of their positive electric charge. The reason is that the forces that bind nuclei together, though very strong, act only at extremely tiny subatomic distances, whereas the positive electric charges of the two nuclei would begin repelling each other at much larger distances. This suggests that the only way to bring about fusion is to create circumstances in which the deuterium and tritium nuclei are moving toward each other at sufficient speeds to overcome their mutual repulsion and allow a close collision, so that nuclear forces can take over. Since motion (or, more precisely, random motion) is heat, fusion is only likely if the deuterium-tritium gas mixture is very hot—at a temperature of many millions of degrees. The recent "cold fusion" attempt to sidestep the need for high temperatures claimed a different mechanism for getting nuclei together, but so far this idea, and almost all other such ideas, have proved to be vain hopes. A genuine exception is "muon catalysis," involving the use of subatomic particles called "muons," artificially created by particle accelerators, to enhance the fusion rate, but this has not yet seemed sufficiently feasible to attract much attention. Thus all serious fusion research today assumes that we must do it Nature's way, by creating on Earth the extremely high temperatures that caused fusion to turn deuterium into helium at the beginning of time.

This much was known in the early 1950s when, flush with the success of having developed the hydrogen bomb, physicists in America, England, and the Soviet Union first began to contemplate how to put fusion to practical use, not as an explosive but as an abundant and—as they thought—environmentally clean source of energy, for the benefit of all humanity. They knew that the idea was an old one. As we have seen, Atkinson and Houtermans had first postulated nuclear fusion as the source of energy in the stars in 1929, a few years before fusion reactions were actually discovered in the laboratory. Researchers in the 1950s knew also that they could not simply copy the hydrogen bomb, in which the high temperatures needed to trigger fusion are obtained by first exploding a smaller fission bomb made of uranium or plutonium. Nor, they thought then, could they simply imitate the fusion that had occurred naturally during the "Big Bang" explosion that created the universe. As it turned out, some fifteen years later scientists would, with the aid of lasers, successfully create in a microscopic pellet of deuterium and tritium conditions very similar to those of the Big Bang. We will return to this "inertial confinement" approach to fusion in chapters 11, 12, and 13. But in the early 1950s, the scientists' thoughts turned instead to the stars.

About a million years after fusion ceased in the early universe, the hydrogen and helium that then constituted all matter had, through the continuing expansion of the universe, cooled to the point that great clouds of these gases began to collect as stars. Because of their great mass, these stars—initially very cold and diffuse balls of gas—began to be compressed under the force of their own gravitational attraction. Any gas cools as it expands and heats as it is compressed. And as the stars were compressed, their centers became hotter, eventually reaching temperatures of millions of degrees, at which point fusion began all over again.

The main fuel for fusion in the stars—which even today is their primary source of heat and light—is the abundant ordinary hydrogen left over from the Creation. (The other main constituent of stars is the primordial helium produced by fusion, accounting for about a quarter of the mass.) The fusion of ordinary hydrogen in stars is a very improbable process, only possible because gravity prevents the escape of the hydrogen so that it eventually undergoes fusion reactions no matter how long it takes. The reaction rate is enhanced by the fact that the density of the compressed stars is high, often much greater than that of water, or any of the other materials familiar to us on Earth. By imagining water heated to a temperature of millions of degrees, one can appreciate the extreme outward pressure in the center of stars, pressure that only stars' enormous gravitational forces could withstand.

In contrast with the exploding early universe, our Sun, or any other star, can be thought of as a steadily burning fusion "furnace" in which the fusion reactions themselves continuously heat the fuel to the extraordinary temperatures required to maintain the fusion process, much as an ordinary fire continually heats and consumes more fuel. Indeed, we often refer to fusion as "burning," and to the fusion reaction, because heat is required, as a "thermonuclear" reaction. The stellar fusion furnace burns more or less steadily because it is held together by its own gravitational attraction, which just balances the tremendous outward pressure of the hot fuel at its core. To imitate a star on Earth, we would strive to create a steadily burning fusion furnace, but one of practical size.

Our first decision would be the choice of fuel, for just as some chemicals burn more easily than others, some types of nuclei fuse more easily than others (and heavy ones do not fuse at all). Scientists in the 1950s soon focused on the DT reaction discussed above. They chose this reaction because the temperature at which fusion proceeds rapidly enough to be useful is lower for this reaction—about 100 million degrees Celsius (180 million degrees Fahrenheit)—than for any other that we might consider. For this same reason, a mixture of deuterium and tritium remains the fuel of choice for the first generation of fusion reactors, although other choices, such as pure deuterium fuel, are still being considered. The use of tritium as fuel does add a complication. Tritium is radioactive, decaying to helium-3 in a dozen years or so, and therefore it is not available in nature. However, enough artificially produced tritium is available for experiments, and the tritium fuel to sustain a fusion electric power industry could readily be produced inside the fusion reactors themselves by bombarding lithium, which is abundant in nature, with the extra neutrons produced in the DT reaction. This added complexity in using DT fuel, and the safety issues posed by the fact that tritium is radioactive, will be discussed in chapters 10 and 15.

Our second decision would concern what kind of force we might apply to hold the hot DT fuel together, as gravity holds stars together. The high temperatures involved preclude simply putting the DT gas in a box. We certainly could not do so if the hot gas were as dense as the interior of stars, for the box would explode. That problem could be overcome by evacuating all the air from the box and introducing only a tiny amount of DT gas, in order to reduce the pressure. But the gas, being in contact with the walls of the box, would quickly cool off and fusion would cease. Still thinking of the stars, the early fusion scientists began to think of ways to suspend the burning gas inside the box, touching nothing, as a star is suspended in space. Realizing that the hot gas would conduct electricity, they began to think of magnetic fields.

Because we will encounter magnetic fields over and over in the next few chapters, let us pause a moment to recall what magnetism is all about. Magnetism was known to the ancients in the form of lodestone, and nearly every child today has played with horseshoe magnets, or the magnets that hold reminder notes on the refrigerator door. The idea of a "field" surrounding a magnet was invented by Michael Faraday and others, in the early nineteenth century, to describe the force between a magnet and a piece of iron (or another magnet). This force is felt long before the objects touch—hence the idea of a field, or aura, around the magnet.

We also speak of electric fields around electric charges. Thus, there is an electric field surrounding every electron and an electric field surrounding the positively charged nucleus of every atom. It is the electric field of the nucleus, reaching out to the electrons, that holds the atom together.

One of the greatest discoveries in all of physics, probably of even greater practical consequence than Isaac Newton's laws of motion, was the discovery that electricity and magnetism are one and the same. From this discovery came the electric generator, the electric motor, the theory of light, the laser, radio, television, microwaves, and the theory of relativity, with its new idea of the equivalence of mass and energy; and from the equivalence of mass and energy came nuclear power and fusion. We will return to this later, when we are better prepared and when we need it. But by way of preview, I would like to share two points that sum it up.

First, there is a fascinating historical linkage between the explanation of magnetism and our quest for fusion today. This linkage started with Michael Faraday's discovery in 1831 of magnetic induction, which launched the age of electricity. The mission of fusion research is the replacement of the polluting fossil fuels that now power Faraday's dynamos by cleaner fusion energy. Yet, as a purely scientific discovery, it was Faraday's law of induction that led to James Clerk Maxwell's theory of electricity, magnetism, and light; then to Hendrik Lorentz's law of motion of charged particles; and on to Einstein's equivalence of mass and energy. These form the foundation of all branches of fusion science today.

Second, as we shall see, the magnetic force itself is truly strange, a challenge for inventors to visualize. Though trained as an engineer and theoretical physicist, I always wanted to be an inventor, and I finally got my chance, in a modest way, through fusion.

All of the pioneers of fusion research, in the early 1950s, were scientist-inventors, working from limited knowledge and keen intuition about magnetic fields. As our first introduction to their accomplishments, let us look at the early work at Los Alamos, spearheaded by the late James Tuck, an English physicist who spent World War II at Los Alamos and later created the Los Alamos fusion program in 1951, at about the same time that Lyman Spitzer began the work at Princeton. The U.S. Atomic Energy Commission gave the secret project the code name Sherwood, which some said was inspired by "Friar Tuck."

The Los Alamos group focused their work on the "pinch" effect, discovered by Willard Bennett in 1934. The first and simplest magnetic pinch experiments evolved from the study of ionized gases, which also gave us fluorescent lights. As we have seen, the atoms of which a gas is composed are usually electrically neutral, with as many negatively charged electrons as there are positive electric charges on the nucleus. However, many circumstances occur in which collisions among the atoms knock off electrons, leaving free electrons, and atoms with a net positive charge, called "ions." (A bare hydrogen or deuterium nucleus, having lost its only electron, is an ion with one unit of positive charge.) For example, in the fluorescent light the electric current creates ions when the streaming electrons carrying the current collide with atoms in the tube. The early universe also was ionized. It was born this way, the temperature being so high that electrons and nuclei could not stick together to form neutral atoms, any atom so formed being quickly reionized by colliding with another nucleus or electron. We call such a fully ionized gas a "plasma." The hot interior of a star is a plasma.

While the vapor in ordinary fluorescent light tubes is only partially ionized, turning up the voltage and current makes the gas hotter, which is to say, its atoms move about more rapidly, until finally collisions among the atoms produce a fully ionized gas, or plasma, similar to the interior of stars (but much less dense). This is the kind of fusion experiment that was first performed, with ordinary hydrogen (not DT fuel), just to see how hot the gas would get. The researchers were also very much interested in the magnetic field created by the currents in the tube.

The voltage applied to a fluorescent light tube, or to the heftier but similar devices used in the early fusion experiments, applies to the negative electrons and positive ions a force that causes them to move in opposite directions along the length of the tube. The electrons, being lightweight, move fastest. The motion of these electrically charged electrons constitutes an electric current flowing along the tube, just as electrons moving in a copper wire constitute an electric current in the wire.

Now, an electric current creates a magnetic field, and magnetic fields exert forces on electric currents. This is the principle of the electric motor. In the pinch experiment, the electric currents flowing in the plasma cause forces that constrict, or "pinch," the plasma, much as gravitational forces constrict or compress stars. However, magnetic forces are much stronger than gravitational forces; hence the hope that strong magnetic fields could be the basis for a practical fusion reactor. Another important difference, not so favorable, is the fact that, whereas gravitational forces act symmetrically, so that stars are spheres, the magnetic force is two-dimensional, acting only in the two directions perpendicular to the direction of the current, so that a magnetically "pinched" plasma is a cylinder.

It is the two-dimensional nature of the magnetic force that makes this force so strange and taxes the intuition of inventors. The question of how to use a two-dimensional force to confine a three-dimensional plasma suggested a variety of arrangements of external magnets to surround the plasma in a satisfactory way. At Princeton, Spitzer's original idea looked like a pretzel, or a figure eight. Appealing to the stars, Spitzer called his device a "stellarator." At the Livermore branch of the University of California's Radiation Laboratory (now the Lawrence Livermore National Laboratory), Richard F. Post tried to block the ends of a linear device with strong magnets that he called "magnetic mirrors." Bending the straight pinch into a circle yielded yet another shape, which, with a note of whimsy, the Los Alamos team called the "Perhapstron," thinking that perhaps it would work, or perhaps it wouldn't.

Though all these inventions had progeny that are still with us, it soon became clear that they tended to have in common two problems that would demand systematic theory and understanding if magnetic confinement fusion was to succeed. These two major problems, soon evident in the early experiments, have dominated magnetic fusion research ever since. First, whereas in stars the plasma pressure and the gravitational forces confining the plasma are in stable balance, a magnetically confined plasma column is often very unstable, the least disturbance causing an initially straight column to kink and bend sufficiently to come into contact with the tube walls. Second, it was generally found that heat energy flowed away rapidly, cooling the plasma.

Fusion researchers struggled in secret for a few years, but in 1958, at the second Atoms for Peace Conference in Geneva, fusion scientists from around the world were allowed to share the results of their research and lay the foundation for one of the most closely collaborative scientific endeavors ever undertaken. This spirit of collaboration evolved none too soon, as the reality of the two major problems — stability and energy confinement — began to turn early optimism into sobering concern.

The first theoretical progress, beginning in the late 1950s, was the development of a principle to determine whether a given magnetic field arrangement would confine a plasma stably. As we have seen, by the time of the Geneva conference the scientists had produced a variety of inventions. But now, with the new "energy principle," at last one could actually calculate which of these schemes would perform stably.

The second major problem, energy confinement, has proved more stubborn and remains at the cutting edge of magnetic fusion research even today. Much progress has been made. A convenient way of measuring this progress is the so-called Lawson number, which indicates how close we are to demonstrating energy confinement sufficient for a practical fusion reactor. The goal—the value of the Lawson number at which energy confinement is adequate for a reactor—is called the Lawson criterion, after J. D. Lawson, who first published it in 1956.*

The idea behind the Lawson criterion is as follows: Consider your kitchen oven. The temperature maintained in the oven depends on both the rate at which electricity (or gas) supplies heat and the rate at which the heat leaks away. If the oven were poorly insulated, the heat would be rapidly conducted to the surrounding air, thereby cooling the oven. In just the same way, the temperature at the center of the Sun, about 15 million degrees Celsius, is determined both by the rate at which fusion in the core produces heat and by the rate at which this heat is conducted outward to the surface of the Sun, where it is radiated in the form of sunlight. The Sun can sustain its fusion reactions in part because it is so large that heat is conducted away slowly. To create a practical fusion reactor, we must compensate for size by using good insulation to prevent rapid heat conduction. Besides confining pressure, the magnetic field also provides this insulation, not only at the edge but throughout the plasma volume, wherever the field penetrates.

We characterize the effectiveness of the magnetic insulation in terms of the "energy confinement time," which is simply the time that would be required for the plasma to cool off if all heating ceased (by convention, it is the time required for the temperature to drop to about one-third its original value). We can characterize the fusion power (the rate of heat production) in terms of the plasma pressure, since higher pressure allows more plasma density, and more density means more fusion power. Also, the temperature must be high enough—about 100 million degrees Celsius for DT fuel; but pressure includes temperature, pressure being the density multiplied by the temperature. Then, as we are using it here, the Lawson number is just the number we obtain by multiplying the plasma pressure by the energy confinement time. When this number is large enough—that is, when it reaches the Lawson criterion—the fusion power can keep the fuel hot enough to burn. It does not matter whether we achieve this criterion by having a very

*The interested reader will find a simple derivation of the Lawson Criterion in the Mathematical Appendix.

large confinement time (excellent insulation) or a very high pressure, or any combination of the two. The number obtained by multiplying the pressure and the time is all that matters.

Steady progress has been made toward meeting the Lawson criterion in magnetic fusion devices. Actually, the Lawson number has roughly doubled every two years for several decades, and there has been about a millionfold improvement since the 1950s. It is this clear and steady progress that has motivated the community of fusion researchers to propose the construction of the world's first experimental fusion reactor on an international basis. This project, the International Thermonuclear Experimental Reactor (ITER), is discussed in chapter 10. Moreover, though less conclusive than the energy principle, theoretical understanding of energy confinement has steadily improved, guided in part by what I will call the "free energy principle," which is similar to the original energy principle but more all-encompassing. We shall return to these two principles in chapter 2.

This was all far in the future when fusion scientists gathered in Culham, England, in 1965, for their second all-fusion international conference since the declassification of fusion research seven years earlier. The air was gloomy with the news that many fusion devices, and especially the stellarator at Princeton, were exhibiting poor energy confinement, the heat leaking at the rate called "Bohm diffusion," which was about the worst that could be expected from the free energy principle, and much too poor to be the basis of a practical reactor. Yet, these were to be the fusion scientists' worst days. As they were to learn at their next conference, in 1968, help was on the way. For the first of several times to come, international cooperation would save the day.

2 : A Formula for Success

. . . .

With appropriate hindsight, we can say that we have known all that was needed to design a magnetic confinement fusion reactor for at least thirty years. The Lawson criterion, which is the yardstick of success, dates from 1956. The energy principle, which evaluates the ability of various magnetic field configurations to confine plasmas stably and is the main tool of fusion reactor design, was first published in 1958. And the first journal papers applying the free energy principle to hot plasmas appeared only a few years later.

In theory, all one had to do was apply the energy principle to select magnetic designs that would confine plasmas stably, and then apply the free energy principle to determine how large the machine had to be to meet the Lawson criterion. In practice, thinking of magnet arrangements (or "geometries") that would stably confine a hot plasma according to the energy principle has proved to be the easier task, while predicting the rate of heat leakage on the basis of the free energy principle is still very challenging. Nonetheless, enough progress has been made in the laboratory to give fusion scientists the confidence to design future experiments, such as the International Thermonuclear Experimental Reactor, on the basis of what we know today. We will look at the reasons for this confidence in the next two chapters and return to the free energy principle in chapter 6. But first let us explore the energy principle, in which there has long been high confidence because much evidence has been found that supports this principle.

Leaping ahead in our story, we will apply the energy principle to what has been the leading design concept in magnetic fusion research for the past twenty-five years, the concept on which ITER is based. This is the tokamak, first discussed in the 1950s by Igor Tamm and Andrei Sakharov in the Soviet Union. The name *tokamak* is a Russian acronym combining the ideas of magnetic field and confinement chamber.

As we had no tokamaks in the United States prior to 1970, my first sight of one was in a laboratory in Moscow in 1964, on my first visit there as the youngest member of the second American fusion delegation to visit the Soviet Union. This was soon after the assassination of President Kennedy. Our hosts were warmly sympathetic, inviting us into their homes. (I remember making paper airplanes, with a white star on one wing and red on the other, for a little boy who had read all of *The Adventures of Huckleberry Finn* in Russian.) The Soviet scientists were eager to show us their experiments, but there was an edge of competition on both sides. I recall using a few Russian words in a speech, partly in response to one of the Soviets, who, when we had met two years before, had goaded me a bit for only knowing English. "Well," I had said, motioning to another colleague from Japan, "you don't know one word of Japanese." "I do," he said. "Tokyo."

I would like to say that I recognized the promise of the tokamak from the start, but such was not the case. In 1966, two years after I first saw a tokamak, I and my co-author failed even to mention the device by name in an article we wrote for *Scientific American.* This was not, I like to think, any prejudice toward things American. It simply indicates the state of confusion then prevailing: no idea showed enough clear promise to warrant special favor. Yet by 1969, though I was then working on magnetic mirror devices, I found myself on a committee recommending the urgent entry of the United States into tokamak research, and by 1987 I was representing the United States in establishing scientific guidelines for the ITER tokamak project.

To introduce the tokamak and the energy principle, let us begin by learning how to design a tokamak together. Specifically, let us learn how to design the ITER tokamak, which is intended to be the world's first experimental fusion reactor and the first tokamak to reach the Lawson criterion. Though it is a bit like tossing someone into the middle of the pool before he or she can swim, I have often given this as the very first homework assignment in my introductory course on fusion, so that my engineering students can discover for themselves what they would need to know in order to design a fusion reactor. Of course, my students do start with a picture of a tokamak, so let us begin with a picture, also.

The tokamak is a cousin of the pinch device we met in chapter 1. As one way to understand the tokamak, recall that the original pinch was a straight cylindrical tube containing a hot plasma. An electric voltage was applied so that current flowed from one end to the other, and this current produced a magnetic field that "pinched," or compressed, the plasma into a column whose hot interior no longer touched the wall of the tube. As noted above, in early experiments with pinch devices it was found that the plasma column confined in this way was often unstable, tending to kink and bend until it moved too close to the nearby tube wall and lost all of its heat. Also, heat flowed rapidly along the length of the column, thereby further cooling the plasma by contact with the walls that sealed off the tube at each end.

Then inventors stepped in. Intuitively, they reasoned that they might reduce the kinking problem in a pinch by adding an external source of magnetic field to "stiffen" the plasma column. They did this by passing a current through an electromagnet called a "solenoid," consisting of wire wrapped, one turn after the other, along the full length of the tube. Further, they could eliminate the leakage of heat from the ends of the tube by bending the tube into a closed circle with no ends. Closing the tube in this way required some new means of driving the plasma current, there now being no ends at which voltage could be applied directly. As we shall see, this problem was solved, at least for as long as was necessary to carry out experiments, by inducing a voltage around the plasma ring, in the same way that an ordinary electric power transformer works. Finally, as its current and temperature increased, the circular pinch would surely expand in diameter, like an automobile tire being filled with air. Though this problem would eventually require an explicit solution, at first it solved itself, since expansion of the plasma current ring automatically induced other currents in the metallic walls of the tube, and these wall currents applied forces that resisted the expansion.

These ideas gave rise to the class of magnetic fusion devices known as "toroidal pinches," meaning pinches in the shape of a doughnut or bagel (mathematically, a "torus"). When the externally produced magnetic field of the solenoid in a toroidal pinch is much stronger than the field produced by the internal current, we call the device a tokamak.

The basic features of a tokamak are shown in the upper part of figure 1. First there is the toroidal tube, or "vacuum vessel," usually made of metal. Air is removed from the vacuum vessel by pumps, and a small amount of deuterium and tritium (or, in most experiments, just deuterium, or ordinary hydrogen) is introduced to create a plasma. A solenoidal or helical winding wrapped continuously around the vacuum chamber provides a strong magnetic field inside the chamber. This is called the toroidal field. Nowadays, the continuous solenoid is replaced by individual circular or elliptical coils placed around the torus. These coils must be large enough to produce the fields required, and the larger the coils, the larger the machine overall. In the first tokamak experiments, such as those I saw in Moscow in 1964, the major radius of the plasma torus was about one meter. (The major radius is labeled R in fig. 1.) The largest tokamaks operating today have a major radius of about three meters, and the tokamak machine with all its parts fills a laboratory space as big as a house. Finally, as already mentioned, there must be some means of generating a current in the plasma. The transformer, or "inductive current drive," method is depicted in figure 1. As in a power transformer, the voltage to drive the plasma current is produced by a primary coil wrapped around an iron core that passes through the plasma and provides a magnetic coupling of the primary to the conducting plasma ring, which serves as the secondary winding of the transformer. A rising current in the primary induces a voltage that drives current in the secondary, the plasma. Modern tokamaks use the same idea, but the iron-core transformer is replaced by a vertical solenoid at the machine axis, which serves as the transformer primary (a so-called air core transformer). The combination of the magnetic field due to the plasma current and the toroidal field produce the pattern of magnetic "field lines" shown in the lower part of figure 1, to which we shall return in chapter 5.

Thus, a tokamak consists mainly of a toroidal tube big enough to hold the plasma that serves as fuel; a solenoidal magnet wrapped around the tube; and a transformer to drive a current in the plasma. But what size tube, how strong a magnetic field, and how much current should be used if the ITER tokamak is to meet the Lawson criterion?

To unravel such problems, the engineering student must learn to focus on the design goal—in this case, meeting the Lawson criterion. To join in the game, the reader should recall that the Lawson criterion involves just two numbers, the plasma pressure and the energy confinement time. The energy principle tells us about the plasma pressure, while the free energy principle concerns the energy confinement time. So, when the students struggling with the ITER homework problem begin to ask me about pressure, I know they are ready to think seriously about the energy principle.

Fortunately for me and my students, the energy principle calculations can now, with fair accuracy, be reduced to two very simple rules. The first rule, which we will call the "pressure rule," says that the more current we have, the higher the plasma pressure we can achieve. This is not surprising, since the tokamak is a cousin to the pinch, and it is the "pinching" of the current that confines pressure in the pinch devices. But the maximum plasma pressure also depends on the strength of the toroidal field and the size of the plasma column. So the pressure rule obtained from the energy principle is as follows: The plasma pressure must not exceed a number found by multiplying the current times the field strength, then dividing by the minor radius of the plasma column (labeled a in the upper part of fig. 1). Pressures higher than this can result in a violent instability, analogous to the blowout that occurs when too much pressure bursts through a weak spot in an automobile tire. Having found that pressure depends on current, my students realize that they must also investigate the limits on current. This leads them to the second design rule obtained from the energy principle—the "current rule." Since the purpose of the toroidal field is to "stiffen" the current-carrying plasma column, we might expect that a stronger field will allow more current, and indeed this is the case. The current rule gives precise meaning to this intuitive idea of "stiffening" the column. It also tells us that, other things being equal, the current can be higher if the column is fatter. What matters is the number we get by multiplying the field strength by the minor radius of the plasma column. To avoid kinking instability, the rule says, the current must not be allowed to exceed this number.

Armed with the pressure rule and the current rule, the students can begin to attack the problem. From the pressure rule, they see that they would like to have the highest magnetic field strength and the highest current possible. According to the current rule, the current can be higher if the field is stronger, so obtaining the strongest possible field helps all around. Therefore, we should design the tokamak magnets to achieve the highest magnetic field strength that we can. Moreover, it turns out that the plasma pressure depends only on the field strength. Even though the two design rules also depend on the size of the plasma column (the minor radius), the minor radius "cancels out" of our calculations because the pressure limit is *divided* by the minor radius while the current limit is *multiplied* by the minor radius, and the final answer is the same for any size tube. The limit on the pressure is simply proportional to the square of the field strength. Doubling the field allows four times the pressure.

While the current rule had been known, more or less, when fusion research was declassified in 1958, it was not until 1984 that the simple form of the pressure rule presented here was first published by Francis Troyon and co-workers, who also provided the numerical factor, called the Troyon number, needed to calculate actual pressures using their formula. The simplicity of Troyon's pressure rule, or formula, came as a pleasant surprise. The most surprising feature concerns the shape of the plasma cross section, which may be circular or elliptical, or whatever we choose. As it turns out, the Troyon number appearing in the formula is the same, whatever this shape. But for a machine of given major radius, we can pack in more current with an elliptical shape, stretched vertically. Thus, by Troyon's formula, a plasma column elliptical in shape can hold more current and more pressure than can a circular plasma column in a machine of the same overall size. As it turns out, the energy confinement time increases also. Troyon's pressure rule summarizes to good approximation the results of numerous exact computer calculations using the energy principle. The rule has been verified in many experiments.

Such strong agreement between theory and experiment is doubly beneficial to progress. First, given the pressure rule, my students now knew with certainty the plasma pressure, one of the two numbers making up the Lawson criterion. Second, the fact that experimental results agreed with Troyon's calculations greatly strengthened the credibility of fundamental theory as a tool for detailed analysis and invention. Today, the energy principle and the theory on which it is based have been incorporated into enormously complex computer programs that reproduce accurately much of the actual behavior of complicated plasmas and guide the designer in optimizing the performance of each new generation of tokamaks.

At this point, our design exercise is half done. We only needed two numbers, the plasma pressure and the energy confinement time, in order to design ITER so as to satisfy the Lawson criterion. From two simple design rules derived from the energy principle, we could find the pressure knowing only the strength of the magnetic field, and this leaves us the freedom to choose the plasma current and the size of the plasma so as to achieve the necessary energy confinement time. We need a new rule, a new formula, to tell us how to do this. Yet even when I began teaching in Berkeley in 1983, I could not give my students a formula to calculate the energy confinement time with much confidence. We had several "guesstimates"— one question, too many answers. This situation soon improved, as we shall see in chapter 3.

Let me add a word about impurities and radiation. As I have described it, the energy confinement time appearing in the Lawson criterion depends only on heat loss via conduction, the direct transmission of heat between objects that are touching each other, such as you experience if you grasp an object hotter than your hand. Plasmas do conduct heat to their surroundings, and it is this conduction process that magnetic fields suppress throughout the plasma volume. But like all hot objects, plasmas also emit radiant heat, on which the magnetic field has no effect. For fusion plasmas, heat is radiated in the form of x-rays, because the temperature is so high. For a plasma of pure deuterium and tritium, this x-ray radiative power is much less than the fusion power that would be generated at a temperature of 100 million degrees Celsius, and therefore we have neglected this radiation in arriving at the Lawson criterion for a magnetic fusion reactor. However, any impurities in the plasma, especially metallic impurities such as might be introduced by erosion of the vacuum vessel, emit more x-rays than would a pure DT plasma. Again, we can generally ignore these matters in establishing the Lawson criterion as long as impurity content in the plasma is kept under control—easy to say, but in fact one of the important technological hurdles that had to be overcome on the long road to the demonstration of fusion energy at Princeton.

Even in the absence of impurities, however, the energy lost via x-ray radiation from a DT plasma always exceeds the portion of the fusion power available to heat the plasma until temperatures exceed about 40 million degrees Celsius, a point stressed in Lawson's original paper. Thus, in order to reach the conditions for ignition, the point at which the plasma burns by itself, some other means of heating the plasma must be provided, at least up to this threshold value of 40 million degrees. Impurities and heat conduction only increase the threshold.

Heat loss by x-ray radiation, being a consequence of collisions of electrons and ions, is unavoidable, as is additional energy loss via the microwave radiation created by electron motion in a magnetic field, though most of the microwaves would be reflected from the vessel walls and reabsorbed by the plasma. However, in this book we will concentrate on heat loss by conduction, which has proved a far more important obstacle to achieving ignition in tokamaks than is radiation. We will return later (in chapters 4 and 9, respectively) to the conditions required for ignition and to plasma heating technology.

3 : A Glimmer of Hope

The laboratory in Moscow which I visited in 1964, which was headed by I. V. Kurchatov, was soon to yield experimental results with far-reaching consequences. It was Kurchatov who, back in 1956, had unilaterally disclosed to Western scientists the Soviet work on fusion research, while the fusion research that was being carried out in England and America was still secret. Kurchatov's disclosure, made during a state visit to England in April 1956, set the stage for the complete unveiling of magnetic fusion research worldwide at the 1958 second Atoms for Peace Conference.

In 1964, the head of fusion research at the Kurchatov Institute was Lev Artsimovich, one of the giants of Soviet fusion science until his death in 1973. Artsimovich placed his faith in a succession of devices based on the tokamak concept. At that time, the popular toroidal configuration in the United States was the stellarator, invented by Lyman Spitzer, who had founded the fusion laboratory at Princeton University to pursue this idea. The original stellarator had been an ellipse twisted into a figure eight. However, the design soon evolved into something that looked very much like a tokamak with no plasma current. In the stellarator, the plasma current was replaced by current carried in external helical windings wrapped around the vacuum vessel. Because it reminded them of the unstable pinch devices, many scientists regarded the plasma current in the tokamak with great suspicion. Perhaps, they reasoned, the stellarator, with no such current, would be better.

It was with great disappointment, therefore, that the Princeton scientists found that heat leaked from the stellarator so fast that experiments intended to achieve a temperature as high as 10 million degrees Celsius actually achieved a temperature only one-tenth that high. Bear in mind that both the tokamak and the stellarator are stable in the sense predicted by the energy principle. The experiments confirmed this expectation, there being in these devices no gross instability of the sort observed in the early pinch experiments. Such violent instability could destroy the plasma in a very short time, between ten and one hundred millionths of a second. However, the experiments showed that, though the plasma was seemingly stable, the heat was escaping in almost as short a time: the characteristic energy confinement time was about a thousandth of a second. Since the energy principle was not at fault, for an explanation one had to turn to additional subtle processes omitted from the energy principle but included in the more comprehensive free energy principle.

In 1965, so little was known about the free energy principle in a hot plasma that one could not explain the Princeton findings on theoretical grounds. True, a variety of instability processes excluded by the simple assumptions of the energy principle had begun to be identified and catalogued by theorists and identified in experiments. Indeed, the seeds of the "right" answer, or *one* of the right answers, had been sown with the theoretical discovery of "drift waves" (which we will encounter again in chapter 7). But the growing list of theoretical instabilities was so long, and the correlation of experiment and theory so unclear, that no one could deny, theoretically, that the persistent, carefully measured rate of heat loss characteristic of the Princeton experiments was a universal, damning feature of toroidal magnetic devices. Mirror devices, also, were experiencing rapid heat and particle losses.

Such was the situation where we left our story at the end of chapter 1, as fusion scientists from around the world were gathering in England for their second international conference, in September 1965. The Princeton scientists duly reported their results that heat seemed to leak from their stellarators at the so-called Bohm rate, a rate that would improve so little with increasing machine size and magnetic field strength that a toroidal fusion reactor of practical size was beyond reach. Yes, it could be true, some theorists said. The new instabilities that they were finding amounted to ways in which plasma electrons and ions could collectively group themselves to create an electrostatic potential difference, or voltage, across microscopic dimensions. Voltages proportional to the plasma temperature, not implausible, would cause ions and electrons to wander out of the machine at just the Bohm rate. That Artsimovich's Soviet team claimed ten times better performance in their latest tokamak was greeted with much skepticism.

The next international conference, which was to be hosted by the Soviet Union, was scheduled for August 1968 at the Academic City established many years earlier near Novosibirsk, in Siberia. A highlight of the meeting was the presentation of results from the latest tokamak, called т-3. Now the Soviet scientists claimed an energy confinement time thirty times that predicted by the Bohm rate, and much higher temperatures.

The new Soviet claims, too bold to ignore, were again doubted at first. At least twice before, fusion scientists had announced startling results only to have to retract them later when better measurements disclosed a faulty interpretation. In two earlier instances, one concerning the Zeta pinch device in the United Kingdom and the other a mirror device at the University of California Radiation Laboratory at Livermore, the problem had concerned researchers' interpreting the production of neutrons as evidence of high temperature. In both cases, the presumption that a few neutrons observed to result from fusion reactions were representative of the average energy of deuterium nuclei proved false. In both cases, occurring within a few years of the declassification of magnetic fusion research in 1958, a healthy international debate over the results had helped to clarify matters. In 1961, it had been Artsimovich who questioned the results from Livermore. In 1968, it was the Soviets' turn to defend their own work against a different view from Princeton.

The debate centered on different ways of measuring the electron temperature. In the small tokamaks then available, the plasma electrons were frequently hotter than the ions, and it is the electron temperature that counts in calculating the Bohm rate of heat loss. If the electron temperature was really what the Soviets claimed, then indeed the energy confinement time must greatly exceed the Bohm-rate prediction; such a high temperature was newsworthy by itself. But if the temperature measurement was in error, all bets were off. The electron temperature reported by Artsimovich and his colleagues was the then-astonishing value of 10 million degrees Celsius, the value Princeton researchers had earlier hoped for but not yet achieved (though, by 1968, they, too, had found some decrease in heat loss compared to the Bohm rate).

Measuring temperatures of millions of degrees is no small challenge. Ordinary ways of measuring temperatures, using thermometers or thermocouples (like that in your kitchen oven), rely on immersing the measuring device in the hot medium itself. This is not possible in a hot plasma, since either the device will destroy the plasma or the plasma will destroy the device. One might try to determine the temperature by observing the radiation emitted from the plasma, since radiation emissions vary according to temperature (as is recognized in the expressions "white hot" and "red hot"). Variations of this idea are actually utilized in plasma research, and in a sense this was the idea behind the neutron measurements ("It was hot enough to produce this many neutrons"). The problem was that a tiny amount of radiation might have other, spurious causes.

The Soviet measurements of temperature in T-3 had been indirect, using special coils designed to detect changes in plasma current. As we shall see later, in chapter 5, in the absence of plasma pressure the externally imposed toroidal field causes the plasma current to twist along helical paths. Heating the plasma, which increases its pressure, causes the current (which must confine the pressure) to straighten out, and this untwisting of the current induces currents detected in the special coils, from which the pressure can be calculated. Since pressure is density multiplied by temperature, given the pressure one can calculate the temperature by dividing by the density measured in some other way. But it was the pressure measurement that Princeton doubted. One spurious effect known to spoil the measurement of pressure by the Soviet method was the presence of a few "runaway" electrons accelerated to high speed by the transformer that produced the current in the tokamak. As we have seen, the purpose of the transformer is to induce around the torus a voltage that accelerates electrons to produce a current. Most of the electrons are only accelerated a certain amount before they collide with an ion, lose their speed, and then accelerate again. Some few, the "runaways," manage to avoid colliding and just keep accelerating to very high energies. The Princeton scientists believed that it was the presence of these "runaway" electrons—not a high temperature for all electrons, that was responsible for the observations.

Within a year, the matter was settled, thanks to what was to be the first of many instances of international cooperation involving the sharing of equipment, as well as expertise and knowledge, within the magnetic fusion community. Remember that this was 1968. Unknown to most of the participants, a major crisis of the cold war had transpired in Prague during the very time of the Novosibirsk meeting. Yet as a result of negotiations between Lev Artsimovich and Bas Pease, director of the Culham Laboratory, by the spring of 1969 a team of British scientists had been airlifted to Moscow. Using a new laser technique available in the West but not at the Kurchatov Institute at that time, the British team set to work remeasuring the temperature in T-3. By August, they were confident enough to report back to Culham, and from Culham to Washington, that the Soviets had been right. The temperature in T-3 was indeed 10 million degrees Celsius. The Soviet measurement of the density was correct, also, and there was no evidence of a runaway electron population in T-3. Thus was born the tokamak revolution, which has made the tokamak the focus of magnetic fusion research ever since, and the standard by which progress on other fusion concepts is measured.

Occurring when it did, the success of the British "airlift" had a very positive effect on the climate for East-West collaboration. Within a few years, bilateral agreements had been drawn up between the Soviet Union and the United States, and also between the Soviet Union and the principal European nations participating in fusion research. These agreements led to frequent exchanges of personnel among the fusion laboratories. In America, the exchanges even included visits of up to three weeks' duration by Soviets to the laboratories at Livermore and Los Alamos, where unclassified fusion research coexisted—separated by fences—with secret work on nuclear weapons.

The British team gave a definitive report of its results at a special fusion conference held at the Soviet accelerator center in Dubna, near Moscow, in September 1969. By then, a tokamak bandwagon was gathering momentum in America, further stimulated by a visit to the Massachusetts Institute of Technology (MIT) by Artsimovich in April 1969. By May 1970, Princeton had converted its largest stellarator to the Symmetric Tokamak (ST); a new tokamak was under construction at the Oak Ridge National Laboratory; and the Atomic Energy Commission was seeking approval for three other tokamaks, at MIT, at the University of Texas, and at General Atomics, in San Diego. Others soon followed in France, Germany, Italy, the Soviet Union itself, and—a bit later—Japan. As it would turn out, this flurry of activity, and the confidence that came with the scientific results it produced, coincided with a new public awareness of the importance of energy, spurred by the oil crisis of 1973. From this came three billion-dollar tokamak machines, one in America, one in Europe, and one in Japan. We will encounter their names many times in the following chapters: in America, the Tokamak Fusion Test Reactor (TFTR); in Europe, the Joint European Torus (JET); and in Japan, the Japan Tokamak 60 (JT-60). These large facilities dominated magnetic fusion research during the past decade, and finally led to the world's first production of "controlled" fusion energy, in JET in 1991, and to a series of world records in TFTR at Princeton, beginning in 1993.

The twenty years between 1973 and 1993 were distinguished, most of all, by a growing confidence that energy confinement in tokamaks would eventually be understood. This confidence was bolstered, first, by the steady advance toward the Lawson criterion, progress only made possible by ever more sophisticated engineering advances as successive experimental devices became larger and more challenging. Confidence grew, also, as the theorists ceased to find so many new instabilities and were able instead to turn their attention to estimating the heat losses implied by these instabilities: in other words, to exploring the free energy principle.

Finally, about the same time that Troyon published his remarkably simple encapsulation of the energy principle in 1984, Robert Goldston noticed that an equally simple law (now known as the Goldston scaling law) may exist for the free energy principle. This law, or formula, gives us a new design rule—the energy confinement rule—which, together with the pressure rule and the current rule, allows us to complete the exercise, begun in chapter 2, of designing ITER so as to achieve the Lawson criterion, the specific value of the Lawson number required for a fusion reactor. According to our new rule, we simply need to design to a specific current, given by the square root of the desired value of the Lawson number (expressed in the right units to give amperes of current). Once the required current is known, we can then use the current rule to find the minor radius of the plasma column (just the current divided by the field strength).

This completes our design exercise. We have seen that, given the magnetic field strength, three simple rules—the pressure rule, the current rule, and the energy confinement rule—determine the plasma pressure, the plasma current, and the size of the plasma required to meet the goals of the design. Even professionals use these approximate rules to choose a design in the right ballpark, and then repeat the design using exact calculations to get it right. The approximate rules are so easy to apply that anyone who remembers a little high-school algebra can do it. In the Mathematical Appendix these rules are set out in algebraic formulas, and numerical information is provided, with guidance, so that the interested reader can carry out his or her own tokamak design.

Whereas both the Troyon formula, or pressure rule, and the current rule were derived theoretically using the energy principle, the energy confinement rule given above, the Goldston scaling law, rested on empirical data organized in a certain way to help us predict the performance of new, usually larger experiments by extrapolating from what we have learned from previous experiments. In other words, whereas the current and pressure rules are reliably based on "first principles," any deviation from these approximate rules being exactly calculable whenever we need it, the energy confinement rule is an "empirical scaling law" and thus could perhaps be violated in an unexpected way.

Some theoretical evidence has been found to support the energy confinement rule (see chapter 7), though for the most part designers still must rely on empirical data, rather than the theoretical free energy principle, to predict the energy confinement time in future devices such as ITER. Moreover, rival scaling rules have been suggested; and even with Goldston scaling the complete scaling formula depends not only on the current but also, in debatable ways, on the plasma size.

Despite their imperfections, scaling laws have served the magnetic fusion community very well. Reliance upon empirical scaling has a venerable history in engineering. To connect this to ordinary experience, suppose that I own a company that makes a profit of $1 million per year. By simple scaling, if you start a similar company ten times the size of mine, you might expect to earn ten times as much, or $10 million per year. On closer inspection, however, you discover that my profit depends on several factors (worker productivity, outlays to suppliers, sales expense, and so on) and that these different factors may not all scale (i.e., vary with size) in the same way. So, to get a better estimate, you ask your company controller to devise a better formula, based on my company's experience, that takes account of these different factors. With these refinements, the controller predicts an annual profit of $8.8 million, and you proceed. The actual profits are $7.0 million in the first year, but as you gain your own experience they rise to $9.2 million in the second year; and you are happy.

So, too, the magnetic fusion community has every right to be happy with its scaling rules—thus far. The step from T-3, the Princeton ST, and other information available when TFTR, in the United States, and the European JET were being designed in the 1970s is enormous, a hundredfold improvement as measured by the Lawson number, more than ten times the extrapolation from JET to ITER. Thus the success of TFTR and JET, reported in chapter 4, is gratifying.

The key to successful scaling is abundant experimental data. It was the agreement of the Goldston scaling law with numerous experiments of different size and character that was convincing. Thus far, an abundance of data about tokamaks has indeed been available, thanks to the worldwide proliferation of tokamaks following the Soviet breakthrough in 1968. However, most of this investment was made in the decade following the oil crisis of

1973, and few new tokamak facilities have been initiated since then. Despite encouraging theoretical progress (see chapters 6 and 7), many fusion scientists fear that progress will cease unless investment in new experimental facilities is resumed soon. We will return to this and other policy issues facing fusion research in chapter 14.

To be useful, experimental data must be both abundant and reliable. What a difference it made when the British team could actually determine, from measurements, whether or not "runaway" electrons were present in the T-3 device. As it turned out later, the Princeton speculation that runaways had confused the Soviet measurements had been correct for another tokamak, TM-3, also reported at the Novosibirsk meeting. That tokamaks could, under certain circumstances, produce runaway electrons was not in dispute. The Soviet achievement in T-3 lay in successfully avoiding those circumstances in order to achieve not merely high energies but energy in the form of random motion, or temperature, as must be the case in a real fusion reactor based on the tokamak concept. If the Soviets themselves had possessed an indisputable means of distinguishing high electron temperature from runaways, there would have been no debate.

As we saw in the case of the temperature measurement in T-3, the problem in obtaining reliable data about what is going on inside a hot plasma is similar to that which your physician faces in trying to diagnose an illness. How can the doctor "see" inside you without harming you? In our lifetimes, new diagnostic tools for doing this—magnetic resonance imaging, CAT scans, PET scans, and ultrasound—have revolutionized medical practice. So, too, the story of fusion includes the development of ever more sophisticated diagnostic tools (see chapters 8 and 12). For example, temperature scans can now be mapped throughout the tokamak plasma with the precision of a hiker's topographic map; and, in inertial confinement fusion, the history of laser-induced microexplosions can be traced over trillionths of a second.

4 : The Real Thing

. . . .

The definitive demonstration of controlled fusion energy on Earth began at Princeton University on the night of December 9, 1993, in the tokamak device called TFTR. After years of experimentation with pure deuterium plasmas, at last the Princeton scientists were ready for the world's first experiments with the 50:50 (half-and-half) mixture of deuterium and tritium fuel required in an actual DT fusion reactor.

The day was doubly significant for the astrophysicist Lyman Spitzer, founder of the Princeton Plasma Physics Laboratory. December 9, 1993, was also the day that NASA astronauts successfully repaired the orbiting Hubble Space Telescope, another "brainchild" of his. "It's wildly improbable," Spitzer said at the time. "It's a remarkable coincidence two projects that I helped get started and did some of the initial, early work on both came to some sort of an important milestone on the same day." Two days later, to mark the TFTR fusion breakthrough, an Op-Ed piece by Spitzer appeared in the *New York Times*. "When I was at the reactor yesterday," he wrote, "I was reminded of how dramatically the program had changed since its inception. In contrast to the beautiful precision of modern engineering, our first experimental device was incredibly primitive. My colleague Martin Schwarzchild and I wound the original copper magnetic-field coils around glass tubes by hand while sitting on the floor of a former chicken barn."

Indeed, the TFTR event was orchestrated with the confidence of a space launch. Reporters were invited, as well as fusion scientists from around the country—myself included, though regrettably I could not make it. According to friends who did attend, the air was filled with hope and excitement. Not a few shed tears of joy. By then, the rival European team at the JET facility in England had already carried out a brief set of experiments using tritium, back in 1991. However, the JET team had deliberately limited the tritium content in the fuel to 10 percent, and that only for a few experimental "shots" of a few seconds' duration, in order to reduce radioactive contamination of the machine. Even these tentative ventures into DT operation had produced over 1 million watts of fusion power, and headlines around the world. Now Princeton was ready for the "real thing," 50:50 DT. The Princeton team had made elaborate preparations to conduct repeated DT experiments safely over a prolonged period of time, thereby fulfilling the ultimate purposes to which the TFTR facility had been committed at its inception, just a few years after the Soviet breakthrough in 1968.

As is typical of large fusion experiments, the Princeton physicists and engineers had worked all day on December 9, 1993, preparing for their first 50:50 DT shots in the evening. Finally, shortly before midnight, soon after the arrival of veteran theorist Marshall Rosenbluth, who had flown in from San Diego, they tried it.

As the magazine *Physics Today* reported the next month, the very first 50:50 DT shot in TFTR generated a peak power of 3 million watts of fusion power, about twice what JET had produced in 1991. By the following night, December 10, the peak power had doubled to a value of 6.2 million watts. And on November 2, 1994, a peak power of 10.7 million watts was achieved, surpassing the 10-million-watt goal established when construction of TFTR began in 1976. A photograph of the inside of the TFTR vacuum vessel is shown in figure 2.

The production of 10 million watts of fusion power, even for the one second available in a single shot in TFTR, is a major milestone on the path to practical fusion energy. Used to generate steam for an electric power plant, the 10 million watts of fusion heat produced by TFTR would have provided electric power for a thousand households, for a second or two. To help us appreciate this achievement, let us again recall the state of fusion research when the TFTR project was first conceived in the early 1970s.

First, in the primitive tokamaks in existence around 1970, the only means of heating the plasma was the current, which produced heat owing to the electrical resistance of the plasma itself, much as resistive wires provide heat in electric appliances such as ovens and toasters. It could be calculated that the plasma resistance, comparable to that of copper in the Soviet T-3 tokamak, would decrease rapidly as the plasma temperature went up, so that this kind of resistive heating was an unlikely candidate to heat tokamaks to 100 million degrees Celsius. Though several alternatives to resistive heating were being developed, all were far from the multimegawatt level that would be required in TFTR. The one chosen, called "neutral-beam injection," was to become a great success, thanks to a heroic development effort by the Lawrence Berkeley Laboratory. Neutral-beam heating (discussed further in chapter 9) has now produced ion temperatures exceeding 300 million degrees Celsius in TFTR, more than thirty times the record-breaking temperature produced in T-3 in 1968.

The second point to appreciate is the enormous step that TFTR represented in plasma energy confinement, compared to T-3 and other tokamaks at that time. Though T-3 already consumed more than one hundred thousand watts of resistive heating power, even with DT fuel it would have produced only a fraction of a watt of fusion power. By contrast, the goal of TFTR was to progress as far as possible toward "break-even"—a fusion power output equal to the heating power. Considering how far one had to go, TFTR came remarkably close to the break-even goal, the neutral-beam power being about 40 million watts to produce 10 million watts of fusion power. The somewhat larger JET facility should come even closer with 50:50 DT operation. This accounting omits the power consumed by the copper magnets in TFTR and JET, since ITER and future reactors will employ "superconducting" magnets that consume negligible power.

Finally, and most important, the DT experiments in TFTR have opened the doors to a new regime of fusion physics, totally inaccessible in earlier tokamaks, in which the byproducts of the fusion process itself play a key role. We call this new field "alpha physics," referring to the high-speed helium nuclei, or "alpha particles," produced in the DT fusion reaction. (The "alpha, beta, gamma" nomenclature used for subatomic particles refers to the order in which various forms of radiation were discovered in the late 1800s, beginning with the "alpha rays" emitted by radium, which were actually helium nuclei; then "beta rays," which were electrons, and so on.)

Understanding alpha physics is crucially important. While neutrons produce most of the energy in a DT fusion reactor, only the alpha particle energy contributes to bringing a fusion reactor to ignition, the point at which the reaction becomes self-sustaining. To understand this, recall that the DT fusion reaction produces fusion energy in the form of a high-speed helium nucleus, or alpha particle, and a fast neutron. Since the neutron has no electric charge, it is negligibly affected by the magnetic field and escapes immediately with all of its energy, making no contribution toward heating the plasma. By contrast, the helium nucleus, or "alpha," containing two positively charged protons, is a charged ion that should be confined by the magnetic field just as the fuel ions are. These are very "energetic" alphas, produced with an energy of 3.5 million electron-volts, which, as random motion, corresponds to a temperature one hundred or more times that of the fuel. Thus, by colliding with the fuel ions, the alphas should heat the fuel, unless the alphas escape too rapidly. Alpha physics pertains to this heating process, and the confinement time of the energetic alpha particles.

From the outset, fusion scientists have been of two minds concerning whether to give higher priority to alpha physics or to the immediately pressing problems of understanding energy confinement in the fuel. For one thing, when significant neutron production was involved there were clearly additional costs associated with radiation protection of personnel and equipment. Proponents of DT experiments in TFTR and JET countered that this drawback was balanced by the technological experience to be gained, and the gain in credibility from dealing directly with the real fusion process. The second point of contention, and ultimately the telling one, was whether or not there was anything special about the energetic alpha component of a DT plasma that would not be learned from experiments with pure deuterium or hydrogen plasmas.

Despite these differences in viewpoint, the decision in favor of DT capability in TFTR and in the European JET had been a relatively easy one in the political climate of the 1970s, influenced by the energy crisis. However, when the funding climate began to change in the United States, around 1980, the tritium-handling capability of TFTR was deferred to a later stage of the project. Thus, though TFTR was completed on schedule in 1982, it still lacked the capability to carry out DT experiments.

The first experiments in TFTR and JET, with pure deuterium plasmas, again focused scientific attention on the energy confinement issue. Ten years earlier, when these machines were being initiated, some theorists were warning that, despite an apparent improvement in performance as the temperature increased in the small tokamaks then available, it was nonetheless to be expected (for reasons we will encounter in chapter 7) that ultimately the energy confinement time might decrease as the temperature increased. The U.S. response was to build another, smaller experiment employing neutral-beam heating in order to explore these issues in parallel with TFTR construction. In 1978, this facility, the Princeton Large Torus (PLT), reached an ion temperature greater than 100 million degrees Celsius, a tokamak record.

I was so excited by the PLT results that I wrote a letter to Jimmy Carter, then president of the United States, to enlist his support for increased funding for fusion research. It occurred to me to do this at a café in Innsbruck, where the PLT results were reported at an international meeting. By the time I reached Paris, the letter had been written and a plan devised to deliver it via a lovely lady back in my hometown in Georgia, also the home state of

President Carter. In due course, the letter was delivered, with a kiss, to the president's budget director, at the Methodist church in LaGrange, Georgia. I do not know if the president ever received it.

Though at first it had seemed that the energy confinement time of ions in PLT was consistent with processes that scale favorably with temperature, by the time TFTR and JET came on line it was becoming clear that the energy confinement time decreased if the neutral-beam injection power was increased in an effort to increase the temperature. Later experiments with other heating methods showed that this was a general result, not peculiar to neutral beams, which could be translated into scaling laws not unlike those suggested by the theorists back in 1973. We have already encountered one such scaling law, the Goldston scaling law, which I incorporated into the energy confinement design rule in chapter 3. Since the rule says that the Lawson number, the single number obtained by multiplying the plasma pressure and the energy confinement time, increases in proportion to the square of the current, it follows that in a machine of given size and current, increasing the temperature and pressure must be at the expense of decreasing the energy confinement time.

Even so, with some getting used to, it was found that the new scaling rule allowed the design of experiments to achieve ignition, and interest in alpha physics focused on obtaining funding for a device in which these experiments could be carried out. However, political changes intervened. In response to lower oil prices and changing priorities, U.S. funding for magnetic fusion research began to tumble, declining from a peak of $471 million in 1984 to about $330 million in 1987. Despite wrenching changes in the U.S. program—the elimination of much work in technology development and almost all research on confinement geometries other than the tokamak—fusion scientists failed to win government support for an ignition experiment. Moreover, keenly aware that an extensive DT experimental campaign would render a fusion device so radioactive that it might no longer be useful for ordinary experiments, neither the TFTR team in the United States nor the JET team in Europe was anxious to commit its one and only facility to DT experiments in the absence of some assurance that a new facility would follow. Thus alpha physics languished on all fronts, while critics inside and outside the program continued to urge that understanding energy confinement was, in any case, the more pressing issue.

In response to this situation, two decisions during 1988–90 set the present course for the U.S. magnetic fusion program. First, to address alpha physics, the Department of Energy decided that, while still advocating an ignition experiment at home, the United States would become a full-fledged partner, at least through the engineering design phase, of the international ITER project. As a corollary, it was also decided that the United States would at last commit itself to carry out the DT experiments in TFTR. Second, to address criticism of excessive reliance on empirical scaling laws for energy confinement, Washington began to shift resources toward experiments, sophisticated measurement techniques, and computer programs aimed at understanding and modeling energy confinement in detail, from first principles. In 1988, to emphasize this change, the Transport Task Force was commissioned by the Department of Energy to coordinate this effort.

The decision to proceed with DT experiments in TFTR was motivated in part by new theoretical concerns that energetic alphas might, indeed, be different. As early as 1978, Marshall Rosenbluth and Paul Rutherford had calculated that any special population of ions at energies much higher than the average could stimulate an instability that might eject these energetic ions from the plasma. Their theory applied to neutral-beam injection, which heats the plasma by depositing energetic ions in the plasma; and also to the energetic alpha particles produced by fusion reactions. If true, this theory cast doubt on the possibility of ignition, which, as we have seen, requires that the alpha particles be confined long enough to collide with fuel ions.

Though neutral-beam heating did work, by the late 1980s neutral-beam injection experiments designed specifically to test the theory were indicating that energetic ions did indeed cause instability, and theoretical refinements continued to point a finger at alphas as well. Predictably, the decision to study alphas in TFTR has stimulated much more theoretical work and further experiments to simulate the instability with neutral beams. While effects initially omitted from the theory and new experimental data suggest that the problem might not be as severe as first thought, alpha-induced instability remains a lively research topic.

The full alpha story is not yet in, and exciting data is still coming in from TFTR. The central issue is, at what density of energetic alpha particles does instability begin? Because a certain amount of time is required for the alphas to collide with and heat the fuel, it must be possible for the alphas being produced by fusion reactions to accumulate for that amount of time if alpha heating is to be efficient. If instability begins to eject the alphas prematurely, only a portion of their energy will serve to heat the fuel. This would in turn require even slower heat leakage from the fuel—or, equivalently, a higher value for the Lawson criterion—if ignition is to be maintained.

Actually, the energetic alpha population in the TFTR experiments is already sufficient for ignition in ITER, with some evidence of instability but nothing disastrous. Indeed, all TFTR parameters, including the DT fuel density and temperature and the density of energetic alphas, more or less duplicate the inner core of an ignited ITER plasma. The extrapolation from TFTR to ITER consists mainly of increasing the minor radius, by a factor of about three, which allows a much higher current in ITER. According to the Goldston empirical scaling and its variations, with a higher current the energy confinement time would be much larger in ITER, which means that the rate of heat leakage from ITER's core would be much less than the rate of leakage from TFTR even though the fuel density, the temperature, and the alpha density are the same. Thus, whereas in TFTR neutral-beam heating is required to achieve this temperature, in ITER itself the alpha heating alone should do the job (i.e., ignition should be achieved).

Meanwhile, stimulated in part in the United States by the redirection of resources in 1988, American fusion scientists and their foreign colleagues have made encouraging progress in looking beyond empirical scaling laws to the underlying fundamental processes. We conclude this chapter with a brief discussion of one such area of progress, the "H-mode."

Whereas the original Goldston scaling for the energy confinement time was based on what is now called the "L-mode" (*L* for "low"), at an international conference in Baltimore in 1982 scientists from the ASDEX facility in Germany first reported a new, improved mode of tokamak operation, called the "H-mode," or "high mode," in which the energy confinement time follows empirical scaling laws similar to those describing the older L-mode experiments but is higher by a factor of two or so. When the ITER project began in 1988, designers were greatly tempted to rely on H-mode scaling as a way of reducing the size of ITER, but despite various speculations, the differences between L-mode and H-mode were not yet understood.

From the first observations in ASDEX, it was realized that the H-mode somehow concerned the edge of the plasma, the transition from L-mode to H-mode being accompanied by a sharp drop in density near the edge. In a good example of the interplay of small experiments, large experiments, and theory, the origin of the H-mode has now been traced with reasonable confidence to the stabilizing effect of electrostatic fields at the plasma edge, discovered first in a small tokamak at the University of California, Los Angeles (UCLA), and later explained theoretically and confirmed on the large DIII-D tokamak at General Atomics, the principal industrial participant in the U.S. fusion research program.

The H-mode story and alpha physics demonstrate why it is so important to understand plasmas in depth, even for the practical purposes of design. Not only do empirically derived scaling laws carry the potential for failure if extrapolated too far, they can also overlook the beneficial effects of plasma properties that are not captured in a few overall parameters such as current, field, and temperature. So it was with the H-mode. As we now understand it, the H-mode is just a way of building a protective "dam" to hinder the outflow of particles and heat at the crumbling edge where the hot fusion plasma joins its cold and hostile environment. Perhaps, with the sophisticated computer programs now being developed (see chapter 7), scientists will be able to discover and utilize such phenomena more speedily in the future.

The H-mode and alpha stories also exemplify the essential connection between theory and experiment that must be maintained if progress is to continue. If experiments are delayed by funding policy, as DT experiments were delayed in TFTR, then theory lags also. I am confident that, if the funding policies of the 1970s had been followed, the achievement of 10 million watts of fusion power in TFTR could have occurred much sooner—perhaps by 1985—and a 1,000-million-watt experimental reactor, something like ITER, would already be a reality. While we cannot revisit these decisions of the past, it is not too late to avoid mistakes in the future.

Many thoughtful scientists believe that it is none too soon to prepare for the energy and environmental problems likely to haunt our grandchildren. In the 1970s, fusion research had already been identified as a vital part of this program of energy preparedness. But today, despite the brilliant success of TFTR, I detect too little excitement among those whose continued support is essential to maintain progress in fusion research. In saying this, I do not fail to appreciate the conflicting priorities faced by public servants who must balance governmental funding of research for the future against immediate societal needs.

Yet, we must not miss the deeper import of the ancient Promethean tale. Besides recounting Zeus's jealous rage when Prometheus gave the Sun's fire to humankind, the ancient Greek authors also recorded, in the names of the characters, a profound lesson in public policy. As I have already noted, the name given to Prometheus, the savior of the human race in the story, means "forethought." There was also another character, his brother Epimetheus, whose well-meaning shortsightedness created the tale's crisis in the first place. Charged to distribute to all the animals those gifts they needed for survival, Epimetheus planned so poorly that he ran out of gifts, leaving nothing for humans. As Prometheus, the savior, means "forethought," Epimetheus, the crisis-maker, means "afterthought."

II

CREATING FUSION SCIENCE

The next few chapters explore more deeply the science and technology of magnetic confinement fusion that led to the actual demonstration of fusion energy, and that laid the foundation for ITER and for the new concepts that are on the horizon. The fundamentals of plasma theory developed in chapter 6 also apply to plasma phenomena in inertial confinement fusion.

Since there is much to cover, I had better say something about where we are heading and what I hope to accomplish in this part of the book. My first objective is to share with the reader what were for me the most exciting ideas to emerge during my nearly forty-year career as a fusion theorist. One such idea is that a magnetic field can be visualized as architecture, a thing of beauty to be shaped to useful ends. This is the theme of chapter 5, which also completes my discussion of the energy principle as a precise design tool now available to the fusion reactor designer.

Another important and exciting idea, which is the main theme of chapter 6, is the strange nature of the magnetic force as manifested in the motion of individual ions and electrons, and the fascinating role that the understanding of magnetism has played in the history of physics. This will lay the groundwork for a discussion of the most detailed version of plasma theory, called the Vlasov equation, so that the reader can better appreciate the ongoing efforts to predict plasma behavior in inertial fusion and to predict the energy confinement time in magnetic fusion devices with the same confidence that we now have in determining the pressure limits using the energy principle.

And a third crucial idea, discussed in chapter 7, is the concept of free energy, as a guidepost toward design features that maximize the energy confinement time and minimize the cost and complexity of magnetic fusion reactors.

My second objective is to share with readers information about some of the technological advances central to progress in fusion research, such as the sophisticated diagnostic instrumentation prerequisite to the understanding of plasma behavior and the validation of plasma theory, and the fantastic engineering advances that make feasible the ITER project (see chapters 8 and 9, respectively). Chapter 10 discusses the ITER project itself.

Especially in the theory chapters, much of the material in this part of the book is necessarily tutorial in nature. However, I hope that the reader who bears with me will find some reward in a glimpse of the pleasure, as well as the complexity, that can be found in the world of plasma physics.

5 : Thinking Like a Magnet

Having covered the main highlights of experimental progress in tokamak research, I now direct the attention of the reader inside the vacuum chamber, to explore further the structure of the magnetic field and the plasma confined by this magnetic structure.

To help us visualize how magnetic fields confine a plasma, I will introduce the idea of magnetic field lines, which are familiar to anyone who has done the experiment of placing iron filings on a piece of paper held above a common horseshoe magnet. If we tap the paper, the filings align in a lovely pattern, as though adhering to "lines" that emerge from the north pole of the magnet and re-enter at the south pole. Actually there are no lines. Rather, each iron chip acts like a tiny compass needle, the imaginary line being the path one would follow in moving from chip to chip, always following the direction of the compass. However, the lines do indicate the magnetic sense of direction, the magnetic force at any point being perpendicular to the field line passing through this point, and they also indicate the strength of the magnetic field, since the lines concentrate where the field is strongest. Thus, physicists and engineers find it useful to describe a magnetic field as a pattern of these imaginary field lines and to describe the field strength in terms of lines per unit of area (e.g., lines per square meter or square inch).

The plasma in a tokamak is confined because, as they move about in the presence of a magnetic field, the electrons and ions that make up the plasma follow the path of the imaginary magnetic field lines. They do this because charged particles move in a magnetic field in such a way that, like the iron filings, they behave like little compass needles pointing the way along field lines. We shall return to this microscopic picture of the plasma in chapters 6 and 7, where we shall see that it is the precisely known motion of charged particles in a magnetic field that accounts in a revealing way for the seemingly bizarre behavior of plasmas that is frequently encountered in magnetic fusion research: for example, current that "magically" springs up all by itself, just where we need it. But this is getting ahead of the story. For now, we need only know that plasma tends to "stick" to magnetic field lines.

Even if the plasma does stick to field lines, it would leak away rapidly unless the field lines themselves remain inside the tokamak. That they do so is shown in the lower part of figure 1. The field lines go round and round the machine, forming fascinating geometrical structures that fit neatly inside the toroidal vacuum chamber of the tokamak. Of course, there is no real structure, only the imaginary magnetic field lines. Thus, to understand how the plasma is confined we must get better acquainted with the magnetic field inside the tokamak.

We often describe the magnetic field of a tokamak as the sum of two fields, one produced by the plasma current and one by the solenoid coils — the toroidal field coil—wrapped around the vacuum vessel (see fig. 1). Magnetic field lines are always perpendicular to the currents that produce them. Thus, the current flowing along the toroidal plasma column produces field lines encircling the column. We call this the poloidal field. And since the current in the solenoid encircles the column, it produces field lines directed along the column. We call this field, with lines traveling round and round the torus, the toroidal field. The actual field is the sum of the toroidal and poloidal components, added in the proper way.

It is this combined magnetic field that results from adding the poloidal and toroidal fields that is depicted in the lower part of figure 1. To construct this field ourselves, imagine that we again use a tiny compass to trace a field line. At any point along the path, we would find that the toroidal field tries to make the compass needle point along the plasma column while the poloidal field tugs sideways. As a result, the compass needle would actually point somewhere between these directions. A little thought will show that following where the compass leads will trace out a field line forming a spiral (or helix), like the stripe on a barber pole.

A friend of mine suggested that one way to make this concept fun would be to model it using a ride at Disneyland. To understand the magnetic field inside a tokamak, you would enter a big circular tube, like the tokamak vacuum vessel. Inside would be a ride something like a roller coaster, but with a track spiraling around and around the machine. Perhaps such a ride would show us what it feels like to be an electron in a tokamak.

Actually, to resemble a tokamak our Disneyland ride would need to have many tracks side by side, and other spiral layers of tracks inside the outer ones. To see the complete picture, we should imagine a circular tube, centered on the toroidal axis of the plasma column, the magnetic field lines being spiral stripes on the surface of this tube. Since the field exists everywhere on the surface, we could use our little compass to trace out a new line wherever we start, and because of toroidal symmetry any pair of lines would travel parallel paths. Thus, by patiently tracing out very many lines, we would find that the spiraling field lines entirely cover the surface of the tube, never crossing. That is, the spiraling field lines themselves construct a surface. We call this a "flux tube," or "flux surface," since field lines are often referred to as "magnetic flux." Now consider other toroidal surfaces inside or outside the first and concentric to it. These surfaces, too, would be covered by spiraling field lines. Thus there is a whole family of magnetic flux tubes, or surfaces, nested one inside the other, as shown in figure 1.

From this visualization of the field inside a tokamak, we can begin to see how magnetic confinement operates. First, if indeed plasma tends to "stick" on field lines, the nested toroidal flux surfaces of a tokamak would appear to be an ideal plasma container with no easy route of escape. Second, we can now see where to look for the weak spot at which too much pressure could cause a "blowout," the possibility of which is included in Troyon's pressure rule derived from the energy principle. To see this, first imagine a flux surface as a straight tube, as it would be in a straight pinch device. By symmetry, the field lines would be equally dense over the surface. But when this flux tube is bent into a circle, the field lines crowd up on the inside and thin out on the outside. The thinning of the lines indicates a weakening of the field on the outside. This is where a "blowout" might occur.

Finally, we can now identify a geometric measure of the stiffening provided by the magnetic field, a visual representation of the prediction made by the current rule obtained from the energy principle. This concerns the pitch of the spiral (analogous to the pitch of a screw). As we have already noted, the spiraling of the field lines is due to the poloidal magnetic field produced by the plasma current. For any given toroidal field strength, the higher the current, the tighter the spiral. Then, "stiffness" turns out to mean that the plasma current should be low enough that no field lines complete even one full poloidal turn in traveling once around the torus. For a typical design, this requires that the poloidal field due to the current be no more than 10 percent of the toroidal field. It is this that sets the limit on current in the current rule. As indicated in the Mathematical Appendix, the full story depends also on the "aspect ratio" (the ratio of major and minor radii), less current being allowed if the aspect ratio is large, implying a longer trip around the torus. The reader may have noticed that there is no guarantee that the spiraling field lines will close on themselves, and indeed on most flux surfaces they do not; they just wind round and round forever, covering the surface but never closing. The special flux surface on which lines do close in one turn around the torus is labeled "$q = 1$" in the jargon of fusion research, where q is the so-called safety factor. Designing so that q is greater than 1 on all flux surfaces, or so that if q equals 1 this happens near the plasma center, is another way to meet the energy principle "stiffness" condition.

In addition to making these qualitative observations, we can now use our mental picture of the tokamak magnetic field to understand a little better what is entailed in actually carrying out the energy principle calculations. The starting point is to find a state of equilibrium, in which all forces are balanced. Otherwise, the plasma would collapse straightaway, like a badly designed building.

We first suppose the plasma pressure to be negligible and consider forces between the plasma current and the toroidal field produced by the solenoidal coil. We have already used several times the fact that magnetic fields produce a force on electric currents. Yet, if the pressure force is zero, in equilibrium the magnetic force must also be zero despite the existence of the current. Now, it is a feature of the peculiar two-dimensional nature of magnetic fields that the force they exert on a current is strongest if the field direction at the location of the current is perpendicular to the current direction, and weaker otherwise, finally becoming zero if the field and current directions are parallel. Thus, for negligible plasma pressure, an equilibrium balance of forces requires the plasma current and the field to be absolutely parallel. Since the field lines spiral, the current paths must spiral also. Then the plasma current will flow partly in the same direction as the current in the solenoid, and therefore the plasma current will change the toroidal field inside the plasma. Working this out, we find that the net toroidal field, including that due to the plasma current, is increased inside the plasma.

The first time I got really excited about magnetic fields was in Tokyo, in 1974, at a lecture on a subject related to the phenomenon described in the last paragraph. The lecture was what we call a "post-deadline" paper—a presentation that the author decided to give at the last minute, the scientific equivalent of a report of late-breaking news. The speaker, Bryan Taylor, a good friend from England, announced that he could explain a curious result seen in the old Zeta toroidal pinch experiment. It appeared that, left to itself, the pinch current would kink unstably as expected, but only until it had twisted itself until the current and field were everywhere parallel, just as I described above. In the course of kinking, the current would distribute itself radially to its liking, then stay that way, perfectly stable. Moreover, Taylor could show all of this from a new form of the energy principle, with the added assumption that a magnetic field property called "helicity" was conserved—that is, constant—during the kinking process. Later we would learn that the constancy of helicity was already known and utilized in astrophysics. But it was news to me.

I must explain that, to a theoretical physicist such as myself, finding a new conservation law (such as the conservation of energy, say) is more exciting than anything else imaginable. Any conservation law is welcome because it provides an enormous amount of information even when detailed calculations are impossible. But a conservation law implying that a plasma could find its own stable state simply amazed and fascinated me. Though it had nothing to do with my own work at the time, I could not stop thinking about it. I am still thinking about it. But more on that later.

To continue our tokamak story: a plasma with negligible pressure is not very interesting, so now let us add plasma pressure to the equilibrium state in such a way that the pressure is higher on the inner flux surfaces (where we expect the temperature to be highest—say, 100 million degrees Celsius) and decreases gradually to a smaller value on the colder, outer flux surfaces (where the temperature is still perhaps 1 million degrees!). This hundredfold difference in temperature and pressure would create a force pushing outward. The only thing to balance this outward force would be the magnetic field, but magnetic fields only exert forces on currents. Therefore, if a balance of forces, including the force due to the pressure difference, is to be possible, then the pressure difference itself must cause an additional current to flow in the plasma. We call this the "diamagnetic current."

Diamagnetism is a basic property of a hot plasma, the one that allows it to be confined by magnetic fields. It differs from the more familiar ferromagnetism of iron, which causes iron to be attracted to a magnet, in that magnets repel, or push on, diamagnetic materials. Thus, if a plasma is diamagnetic, we can hope to confine it by surrounding it with magnets pushing in from all sides. Of course, the magnets must be arranged very carefully to prevent any route of escape, a task made more difficult by the fact that the plasma is not a solid but a gas, a "diamagnetic fluid."

Just as ferromagnetism can be traced to the behavior of tiny electric currents in individual atoms, so also plasma diamagnetism can be traced to the motion of individual ions and electrons in response to a magnetic field, these motions again constituting tiny currents, since current is simply the motion of charges. However, for many purposes it is unnecessary to consider these tiny currents in detail. Rather, pursuing the fluid analogy, we can represent the diamagnetism of the plasma by the additional current that must flow in the plasma to maintain force balance if there is a pressure difference. In mathematical terms, we can simply add to the force balance equation a term proportional to the pressure difference. We would then find that, if the force balance is maintained, the current and the magnetic flux lines are no longer exactly parallel, and the extra current that makes this so is just the diamagnetic current due to the pressure difference.

This way of describing plasmas—representing them as an electrically conducting, diamagnetic fluid—is called "magnetohydrodynamics," or simply MHD. Fusion scientists inherited MHD theory from astrophysics. In 1970, the Swedish astrophysicist Hannes Alfvén, who has made major contributions to fusion research, was awarded a Nobel prize for his part in developing MHD theory. Lucky for me, before his prize was announced I had by chance already invited Alfvén to be the banquet speaker at a national meeting of plasma physicists in Washington. So I was able to introduce him as the first Nobel laureate in plasma physics.

The energy principle, published in 1958 by Ira Bernstein, Edward Frieman, Martin Kruskal, and Russell Kulsrud, at Princeton University, was based on the MHD force balance equation. Shortly thereafter it was shown to be compatible with the Vlasov equation, the more comprehensive theory that we shall encounter in discussing the free energy principle in the next chapter.

The idea behind the energy principle is easily understood from a mechanical analogy. Consider a marble at rest at the bottom of a cup sitting on a table. In this position, all forces are in balance, the pull of gravity on the marble being opposed by the force exerted by the cup (and the force on the cup being opposed by the force exerted by the table, the force on the table being opposed by the force exerted by the floor, and so on). If I nudge the marble up the side of the cup, gravity would exert an additional force pushing the marble back toward the bottom of the cup. Even if I let go, the marble would only roll back and forth near the bottom, never escaping. That is, the force balance is stable, since a little disturbance hardly changes anything. Now I turn the cup upside down. With patience, I could still balance the marble on top, but now even a tiny disturbance would cause it to fall. This is an unstable situation.

Even if I could not see the marble and cup, I could distinguish stable from unstable arrangements in a simple way if I were able to measure the way in which the combination of the cup and gravity exerted forces on the marble. First, I would look for "equilibrium," the condition in which the forces were just in balance. For the marble-and-cup problem, I would find just two equilibrium situations, the two described above. To test for stability, I would calculate the net change in force when the marble is displaced from these equilibrium positions. Equivalently, I could calculate the work I must do to move the ball with my finger. As the term is used in a scientific sense, "work" is the energy I would expend. This energy is equal to the force required to move the marble multiplied by the distance the marble moves. To test stability, I would calculate the work, or energy, required to move the marble in various directions. For a marble in a "perfect" cup, one with no irregularities, this would be the same in all directions, but what if there were a hole in the cup? Remember, I cannot see the cup, though I do know the force it exerts.

From this example, we can see how the energy principle works. First consider the cup in the upright position. Then (if there were no holes!) my calculation would show that it did take work to move the marble, which is why this is a stable situation. Whichever way I moved the marble, I would calculate that a positive energy (or work) was required to move it. But when the cup was turned upside down, my calculations would only give negative values for the work, meaning that no work was required to move the marble. In fact, the marble would gain energy, from gravity, by moving. Hence, the situation is unstable. Of course, if the cup had a hole, I would have discovered it. With the cup right-side up, I would have found one displacement (toward the hole) that required no work.

In just the same way, to use the energy principle we must first look for equilibrium situations and then examine deviations from them to test for stability. Because the plasma is a fluid with no internal strength (i.e., it tends to flow in response to any applied force, unless it is confined), this requires carefully checking the force balance at every position within the plasma. The MHD force balance equation allows us to do this in a systematic way, point by point. Then, to test for stability, we must determine every possible distortion of the plasma away from its equilibrium shape. Next we must calculate the work done to create the distortion at every point in the plasma, and add up all these quantities of work (or energy). If the total work is positive for all distortions, the system is stable. Then, even if distortion would cause some portion of the plasma actually to gain energy, these gains would be more than offset by the total work required. Conversely, if the total work was negative for some distortion, the system would be unstable, meaning that it could distort in this way all by itself. It was just such elaborate calculations applying the energy principle to many, many cases that produced the pressure and current rules discussed in chapter 2.

While the energy principle provides a precise method for calculating stability, it is very helpful to have a physical picture, also. Returning to our mental image of plasma stuck on magnetic flux surfaces, we see that equilibrium is just uniform pressure on each surface, since plasma flow along field lines would equalize the pressure over each flux surface. Since a drop in pressure exerts an outward force, even in equilibrium the outer flux surfaces would be distorted somewhat by the high pressure on the inner surfaces. Finally, since the plasma is stuck on the lines, distorting the plasma also distorts the field lines and hence the magnetic field strength. The magnetic field is a form of energy, the energy density being proportional to the square of the field strength. Hence distorting the plasma, which distorts the field, may increase or decrease the field energy, depending on circumstances. The field outside the plasma column, partly due to the plasma current, is also distorted. And distorting the plasma may compress it (which increases its energy) or allow it to expand (which decreases its energy), all such effects being included in the energy principle calculations.

From such mental pictures, one gains insight that is helpful in sorting out promising ideas. Thus, one can easily see why the solenoid field might stabilize the pinch. Without this external field, the column of circular field lines produced by the plasma current would be easily bent, like a flexible spring; the straight field lines of the solenoid, however, stretched by bending, would resist being bent and would thereby stiffen the column.

Let me give one more example of design aided by visualization, which also provided an early test of the energy principle. It involves the "magnetic well," which relies only on the diamagnetic nature of a plasma in order to confine it (whereas the tokamak depends on both diamagnetism, to confine the pressure, and good electrical conductivity parallel to the field lines, to carry the current). The magnetic well is an improved version of the magnetic mirror concept mentioned in chapter 1. The simplest magnetic mirror device consists of just two circular magnet coils facing each other, separated by a distance somewhat greater than the coil diameter. The magnetic field strength is large in the vicinity of each coil but weaker midway between. A hot plasma cloud located at this midway point, being diamagnetic, is "reflected," or repelled, by both coils (acting like "mirrors"), thereby achieving a force balance. Magnetic mirror reflection only works if the pressure is weaker parallel to field lines than it is perpendicular to the field, a condition that is ultimately a cause for leakage. Incidentally, the plasma within the ionosphere high above the Earth is contained there by magnetic mirror reflection created by the Earth's magnetic poles. It is the leakage of this plasma that causes the aurora borealis.

The simple two-coil mirror system is unstable, since the repulsive force of the coils only serves to center the plasma if the plasma is located directly between the coils. A sideways displacement away from the equilibrium position causes both mirror coils to push the diamagnetic plasma further in the sideways direction, and the bigger the displacement the stronger the push. Applying the energy principle, we would find that the work done by any sideways displacement is negative; the displacement takes no work. However, the addition of other magnets surrounding the original two might push the plasma back, thereby restoring stability. This is the idea of the magnetic well.

Experiments to test the magnetic well idea were carried out in Moscow by Mikhail Ioffe and co-workers, culminating in 1963. In these experiments, the magnetic well was created by an array of conducting bars forming a cage around the two circular coils. Calculations showed that when current was passed through these bars (alternating in direction from bar to bar), as few as four bars, together with the circular coils, would produce a magnetic field that is stronger with increasing distance, in every direction, from the equilibrium position. Thus, no matter which way the diamagnetic plasma moved, the field would repel it back toward the equilibrium position. Any displacement would therefore require work, so the energy principle predicted stability, and the experiments of Ioffe and co-workers, initially with six bars, proved that a plasma confined in this way was in fact stable.

Ioffe's experimental demonstration of magnetic well stabilization produced great excitement. Many things could have gone wrong, ranging from useless stabilization effects special to the low-temperature plasmas then available, to some dreaded new instability to baffle theoreticians once again. That it worked was unmistakable, the stabilizing effect being clearly present only when the extra conductors to provide a "well" were energized. To honor the event, we in the West quickly dubbed these conductors "Ioffe bars" (though it took thirty years of political change to allow Ioffe to be honored formally in America).

While the Ioffe bar solution that could be turned on and off was ideal for the first experiment, later applications prompted an improved design, derivable from the four-bar case but involving only a single coil twisted into the shape of the stitching on a baseball (or a tennis ball, as the English co-inventors preferred to say). The pattern of magnetic field lines produced by a baseball coil is sculpturally beautiful, a twisted fan of parallel lines that could hardly have escaped the notice of artists. Indeed, some fifteen years after Ioffe's experiment, I was walking down a street in Atlanta when, to my amazement, I saw this very magnetic field pattern modeled in gleaming strips of chrome, in front of a bank. I later learned that the sculptor, Charles Perry, had been equally surprised when, back in 1966, he had found a sketch of his sculpture—actually, our magnetic field pattern—in the *Scientific American* article I had co-authored. When Perry and I finally met in person, in 1991, he explained how he, too, had derived the idea for his sculpture from the stitching on a baseball, visually cutting and folding along the seam until he saw, in his mind's eye, the same lovely shape that our magnet produced as an equally imaginary pattern of field lines.

6 : Heat Death and Relativity

· ·

The MHD theory and the energy principle are major accomplishments of fusion theorists, providing as they do a reliable means of judging the stability of plasma equilibria in magnetic confinement devices and calculating the allowed pressure that goes into the Lawson criterion. In this chapter, we will dig a little deeper into the foundations, beyond the MHD idealizations of perfect electric conductivity and diamagnetism, to consider the individual particle motions and turbulence that must be taken into account in calculating the other half of the Lawson criterion, the energy confinement time. For this we must touch upon the two most profound accomplishments of the golden age of classical physics. One is Maxwell's theory connecting electricity and magnetism, which led to a new understanding of the magnetic force known since ancient times, and which led eventually to Einstein's theory of relativity. The other is the second law of thermodynamics, the basis for our free energy principle. This is the law, now thought to apply to all branches of science, that tells us that "order" always turns into "disorder," meaning that a unique object such as a magnetically confined plasma will meld into and become more like its surroundings as time passes. It was this recognition that all highly organized things tend toward a disordered state of dullness that inspired the nineteenth-century notion of the ultimate "heat death" of the universe. It also inspired fusion skeptics to predict that magnetically confined plasmas would meet a similar fate, much more quickly.

The topics covered in this chapter lie close to my own introduction to plasma theory, in my first job at the Oak Ridge National Laboratory back in 1957. In those days, none of us had been taught plasma physics in school, so everything had to be learned on the job according to the needs of whatever program had hired us. Thus, while my colleagues at Princeton were perfecting MHD theory and the energy principle for stellarators, I was learning about the motion of ions in magnetic mirror machines.

Going in Circles

The motion of an ion or electron in a magnetic field brings us face to face with the strange nature of the magnetic force to which I have alluded throughout the book. The facts are simple enough. An ion moving perpendicular to a magnetic field line merely moves in a circle. An electron in the same circumstance also moves in a circle, but in the opposite sense: if the ion moves clockwise, the electron moves counterclockwise. This is in fact a commonplace thing. It is the circling motion of electrons in a magnetic field that causes them to emit microwaves in your microwave oven.

The strange part is this: thinking about other kinds of forces such as gravity or electric charge, one might conclude that if an object moves in a circle there must be something at the center of the circle holding on to it, as the Earth's gravity holds onto the orbiting Moon. Yet, an ion (or electron) moving in a magnetic field encircles nothing at all! And if I could gently move the ion over, it would happily continue moving in a circle at its new location, still going round and round with nothing to hold it there.

Well, you might say, the magnetic field holds it. And if you studied physics in high school, it might occur to you that the moving ion is like a current, since electric current is just charge in motion, and magnetic fields do act on currents. Going a little further, you might recall that the magnetic field pushes sideways on a current that is perpendicular to a magnetic field line. Then the light dawns. If the perpendicularly moving ion is like a current, and the field pushes sideways, it is clear why the ion turns in a circle. It is as if I were driving a car crosswise to a fierce gale. As I go forward, the gale blows me sideways and I begin to turn. If, like the magnetic force, the direction of the gale turns with me, always hitting me broadside, I keep turning and turning until my path closes in a circle.

This is one way to explain the ion motion, and it is the actual approach we will take to calculate it. Having a moment of insight like this is a good feeling. Theorists especially thrive on such rare moments of revelation when, after months or years of puzzlement, you say, so that's how it works! But are we not sidestepping the question? It is still true that the ion encircles nothing, for the magnetic field is not as tangible as the Earth, and the field exists everywhere, not merely at the center of the orbit. And anyhow, what kind of force depends on motion? Isn't motion a relative thing?

Having studied theoretical physics in graduate school, I more or less knew the answer to these questions as I first began studying the problem, when I was new on the job back in 1957, but really coming to understand it gave me a lot of pleasure, especially as I began to put together the history behind it.

We could say that the "real" answer is that magnetism is a trick of Einstein's relativity, just electricity in disguise. In ordinary experience the magnetic force is so weak that we would never notice it except for the fact that the electrostatic force itself is so strong that Nature nearly always contrives to hide it. To understand this, suppose that the magnetic field is created by an electric current in a big coil, like the coils in a tokamak. The existence of the current means that electrons are streaming through the metallic conductor in the coil. The negative charge of each of these electrons should attract our circling ion, along a line of sight between the ion and the electron, but actually the spinning ion does not experience these forces because Nature has made sure that the positive charges of all the other ions in the metal coil just equal the negative charges of the electrons; the coil has no net charge (usually). Then the repulsive electrostatic force of all these metal ions acting on our circling ion just balances, or "cancels," the electrostatic attraction by the electrons. So why is there a "magnetic" force?

The answer, worked out at the turn of the century by the Dutch physicist Hendrik Lorentz, goes to the heart of one of the great discoveries of nineteenth-century physics, the discovery that electricity and magnetism have a common origin and ought to be regarded as a single "electromagnetic" force. The first clue was the discovery by Hans Christian Oersted in 1820, furthered by André Marie Ampère, that an electric current produces a magnetic field. This was shown quite simply by demonstrating that passing current through a wire attracts the needle of a compass.

Then came Michael Faraday's 1831 discovery of the law of induction, which led to the invention of the electric dynamo, or generator, which made possible a world powered by electricity as we know it today. Faraday and others had reasoned that if electric current could produce magnetism, then magnetism ought to produce electricity, and they had tried unsuccessfully to prove it until, with characteristic attention to experimental details, Faraday discovered that it was not the magnetic field per se but a *changing* magnetic field that did the trick. Looking back, this all makes sense, of course, since electric power is a source of energy, and we ought to expect to have to do work (that is, *change* something) to generate this energy, as steam or water power actually must do to run a generator.

By 1864 the Scottish theoretician James Clerk Maxwell had put this all together in the equations known today as Maxwell's equations, which we will use to calculate the electric and magnetic fields due to plasma charge and current. In combining Ampère's law and Faraday's law of induction, Maxwell found that, just as Faraday's law required the rate of change of the magnetic field, logical consistency also required adding the rate of change of the electric field to Ampère's law. This addition turned Maxwell's equations into a theory of light, and led to the invention of radio and television, which have revolutionized world communication. Solving these equations showed that, oddly enough, light travels at a definite speed, even if its source is moving. This was regarded as odd because, by analogy with sound waves in air, one would expect the speed of light waves to appear greater if the source of the light moves toward you, the observed speed being some fixed light propagation speed relative to its source, added to the speed of the light source itself. Adding speeds in this way was what the term "relativity" meant at that time, a meaning that dated back to the mechanics of Newton and Galileo. Then, when experiments showed the observed light speed to be constant after all, in 1905 Einstein turned everything upside down, keeping Maxwell's equations and amending Newton's, to produce his Special Theory of Relativity, which eventually led to the prediction and discovery of fusion energy.

Lorentz made two key contributions to these developments. One was his somewhat contrived attempt to reconcile the observed constant speed of light with the old ideas assuming the existence of an "ether" to propagate light as air propagates sound. This led to the Lorentz "transformation" equations, mentioned below. Later Einstein abandoned the hypothetical ether but kept the equations. The other contribution was the Lorentz force, which acts on a charged particle in the presence of electric and magnetic fields. The Lorentz force law simply says that the Lorentz force is the sum of the electric force and the magnetic force "in any chosen reference frame."

In practice, this is usually the "laboratory frame," the place where I am standing when I measure the electric field and magnetic field in order to calculate the combined force on the particle. To calculate the magnetic field's contribution to the total force, I also must know the velocity of the particle. Then the Lorentz law says that I must measure all three quantities—the electric field, the magnetic field, and the particle velocity—from the same vantage point, or "reference frame." This is because all three quantities would appear to be different if I and my equipment were mounted on a cart moving through the room. For example, if the cart could move as fast as the charged particle, I would find no magnetic force on the particle, since it would appear to be at rest if I moved beside it. But now I would measure a stronger electric field to compensate for the missing magnetic force. In other words, electric and magnetic fields are one and the same thing, interchangeable depending on the state of motion of the observer, according to well-prescribed rules. These rules, called the Lorentz transformation, are the essence of Einstein's theory, worked out earlier by Lorentz, though it remained for Einstein to enforce these rules upon Newton's mechanics, which is possible only if Newton's immutable mass actually changes as the energy changes. Hence the equivalence of mass and energy, $E = mc^2$.

Suppose I decided to apply Lorentz's force law to calculate the contribution of a particular electron (strictly speaking, an electron moving at a constant speed, in a constant direction) to the fields while I was moving with that electron. Then I would find only an electrostatic field in the neighborhood of the electron, the kind of electric field that one could produce by rubbing a pocket comb or a piece of amber. Yet, if I were standing still, I would say that, since the electron is moving, it also produces a magnetic field. Lorentz's transformation law tells me what magnetic field I should expect to see. If the electron speed in the laboratory frame was very small by comparison with the speed of light, the magnetic field would be much weaker than the electrostatic field, which would be only slightly less than what we measured moving with the electron. That is, at low speeds, the magnetic field is only a small correction to the electrostatic field, due to the motion.

The relativistic origin of the magnetic field and the force it produces on a moving charge is given away by the appearance of the speed of light in the answer, though most physics texts would call the magnetic force "classical." Thus, though Einstein's special theory of relativity only came along in 1905, the ancient Greeks examining lodestone were merely playing with a weak relativistic correction to their other hobby of charging up amber.

As in our example of a coil producing a magnetic field, our interest in the weak magnetic force in plasma physics lies in the fact that, in the plasma as in the coil, the negative and positive charges are about equal in number, so that their electrostatic forces cancel each other out. But due to the magnetic correction, the cancellation is not complete, the "leftovers" being the magnetic force. With this new understanding of the origin of the magnetic force, we can now begin to see how it might be so peculiar—peculiar enough to cause an ion to move in circles, encircling nothing. Actually, the force between a pair of charged particles at rest is directed along the line of sight between them, just as we would expect by analogy with the gravitational force. But if the particles are moving, the direction of the force is generally no longer exactly along the line of sight but is slightly askew, by a small angle that depends on the speeds of the particles divided by the much larger speed of light. If I were to focus on one of these particles as a "test charge," when I added up all the forces acting on it due to all other moving charges producing currents in a coil or currents in an overall charge-neutral plasma I would find that all of the line-of-sight electrostatic parts canceled, but the little skewed deflections due to the relativistic corrections to the force do not cancel: they add. All of these deflections would turn out to be perpendicular to the direction of motion of the test charge, just as we would expect of a magnetic force. Indeed, the sum total of these deflections *is* the "magnetic force."

Putting it all together, then, a moving ion travels in a circle not because there is something to hold it there, but rather because of all of the little sideways kicks due to the not-quite-cancellation of the not-quite-line-of-sight forces acting on it from other moving charges, some moving in coils relatively far away.

While in most engineering applications one need not be concerned with charges at the atomic level, in plasma physics we really do, in one way or another, compute the fields by adding up the charge and current due to each of the ions and electrons in the plasma. The procedure for doing this is embodied in the Vlasov equation, which is our main theoretical tool for going beyond the MHD theory, in order to calculate the energy confinement time from the second law of thermodynamics and the free energy principle.

Entropy and Turbulence

To apply thermodynamics to the problem of heat confinement in plasmas, we introduce the concept of "entropy," which is a number that gives precise meaning to "disorder." The second law of thermodynamics states that the entropy of a system never decreases and tends to increase with time. The rate of entropy increase is related to the rate of heat loss. Identifying and calculating the processes that cause the entropy to grow is the hard part.

The first precise calculation of the rate of entropy increase in a tokamak using the Vlasov equation, and thus far the only complete theory of this kind, is the work of Marshall Rosenbluth, Richard Hazeltine, and Frederick Hinton, published in 1972, based on the "neoclassical theory" of Roald Sagdeev and Albert A. Galeev in the Soviet Union. It is this neoclassical plasma theory, discussed in chapter 7, that predicts the strange behavior I alluded to earlier—the self-generated "bootstrap" current, now confirmed by experiment and put to use in design.

In the neoclassical theory, the processes increasing the plasma entropy are the known ones, namely, collisions among the ions and electrons. For example, the current in a tokamak is a form of "order," the existence of a current implying an orderly flow of the electrons relative to the ions (since current is charge in motion). Collisions between electrons and ions represent a friction, or resistance, that would eventually slow down the electron flow and destroy the current, leading to a state of less order, or higher entropy. When the current is gone, plasma confinement is gone.

A collision between an electron and an ion can be thought of as a momentary departure from, or fluctuation in, the careful balance of forces that constitutes the equilibrium state. The interesting situations in complex systems are often those that, in an average sense, repeat themselves over and over in most details. When average properties such as density are constant in time, we say that the system is in a state of equilibrium, which, in a plasma, further implies almost equal ion and electron densities, smoothly distributed, producing little or no electric field. When the equilibrium state changes little even if disturbed, we call it a stable equilibrium. But even in a stable plasma equilibrium, due to their random motion two electrons may have a close encounter that causes the electron density to increase locally, with no corresponding increase in ion density, thereby producing a momentary fluctuation in the electric field (as would a close encounter of an ion pair, or an electron-ion pair). And this random fluctuation in the electric field in turn affects the motion of the electrons. Similarly, the acceleration or deceleration of one electron encountering another causes a fluctuation in the local current and magnetic field, again affecting the particle motion. It is the effect on particle motion due to the randomly fluctuating electric and magnetic fields arising from close encounters that destroys the orderly current flows and increases the entropy. The weaker the fluctuation effects, the slower the increase in entropy and the longer the energy confinement time.

Collisions such as those described above—close encounters of charged particles two at a time—are not the only or even the most important source of fluctuations that cause the entropy to increase. It is the nature of the force between two charges that collision rates decrease as the temperature increases. Thus, the collisional increase of entropy included in neoclassical theory should greatly diminish at high temperature. However, as we have seen, according to empirical evidence, such as the Goldston scaling for the energy confinement time, discussed in chapter 3, the rate of entropy increase and heat leakage actually increases with temperature, in total disagreement with the neoclassical prediction. This excessive heat leakage, "anomalous" compared to neoclassical expectations, is generally attributed to instabilities on a microscopic scale. These microinstabilities produce measurable fluctuations of the electric and magnetic fields of a very different, "collective" kind, involving large groups of plasma particles acting together.

Collective motion is the most important characteristic of plasmas. We can more or less categorize collective motion as either orderly and coherent, or random and turbulent. Only the turbulent variety contributes to the relatively slow increase of entropy that we have been referring to above, while the coherent collective motion is either stable and benign or violently destructive, like the kink instability in early pinches, which simply must be avoided. It is this approximate separation into coherent and turbulent collective phenomena that has motivated our distinction between the energy principle, based on MHD theory, and the more comprehensive free energy principle, which we shall apply only to systems already known to be stable on the basis of the energy principle. I should hasten to admit that some of my colleagues would object to such a simple categorization of turbulence into just two types. But then, as the saying goes, there are two types of people in this world: those who divide everything into two types, and those who don't.

The simplifying features of MHD theory that bring out the potential for coherent collective motion are the assumptions of perfect diamagnetism and perfect conductivity that cause the plasma to "stick" to the magnetic lines, as assumed in chapter 5. If the plasma sticks to field lines, the field lines move only if the plasma moves. Since no piece of the physical plasma can instantly leap to another location, so also no piece of an imaginary field line stuck to the plasma can make an instantaneous change. In other words, in MHD theory, field lines can bend or twist but they cannot break. Applied to tokamaks, this means that the orderly structure of nested magnetic flux surfaces (shown in fig. 1) can twist and distort coherently, but the lines cannot break to form a radically different structure.

Only distortions of the coherent kind allowed by MHD theory were considered in applying the energy principle in chapter 5. In the stable case, analogous to the plucking of a guitar string, the "plucking" of magnetic lines merely produces wavelike vibrations whose frequency and wavelength are determined by the inertia of the field lines due to the plasma mass stuck to the lines, and by the force or "springiness" of the field lines due to changes in magnetic field energy and plasma pressure produced by the distortion. In the unstable case, the distortion grows rapidly, as does the likelihood that the plasma will crash into the vacuum vessel and be destroyed. This way of destroying the plasma also increases the entropy, but not in the slow, incremental manner that we have in mind when we speak of heat leakage.

The MHD theory is an approximate picture, only as accurate as its underlying assumptions of perfect diamagnetism and perfect conductivity. But we can always test its validity against a more complete theory, and when we do, we find that MHD theory is usually valid except for distortions on a spatial scale much smaller than the dimensions of the plasma. It is these exceptions that give rise to the "microturbulence" that increases the entropy only slowly if the system is MHD-stable by the energy principle. Thus, to deal with heat confinement and the increase of entropy, we must learn something about this more complete theory, the Vlasov equation, which was first published by A. A. Vlasov in 1945.

Validated by numerous experiments, the Vlasov equation has long been accepted as the "first principles" theory of collective behavior in plasmas. The MHD theory can be derived as an approximation to the Vlasov equation. But the Vlasov equation also covers cases in which the simplifying assumptions of MHD theory (perfect diamagnetism and perfect conductivity) are no longer valid. It is the failure of these assumptions that gives rise to collective microturbulence. Because of the Vlasov equation's central role as the arbiter of credibility in plasma theory, we shall devote the remainder of this chapter to understanding it, and its relation to MHD theory, in some detail. Only then will we be prepared, in chapter 7, to see that understanding entropy, free energy, and heat confinement may not be so difficult.

I first learned about the Vlasov equation in 1957 from unpublished notes on plasma physics that had been prepared by Edward Harris, one of my colleagues at Oak Ridge. Harris writes and thinks about physics more simply and clearly than anyone else I have ever known, and he went on to become a professor of physics at the University of Tennessee and author of a wonderful two-volume textbook covering all of theoretical physics. In 1963, relying heavily on Harris's work, I prepared my own set of unpublished notes, which, in 1973, Nick Krall and Al Trivelpiece included, in part, in their still-classic textbook *Principles of Plasma Physics*. So Harris's notes, so useful in my own education, were finally published after all.

Whereas MHD theory begins with familiar concepts such as density and temperature, the Vlasov theory begins with the motion of individual ions and electrons, calculated from the Lorentz force. As we have seen, according to Lorentz, an ion (or electron) attempting to move perpendicularly to a magnetic field line simply moves in a circle—one of small diameter if the magnetic field strength is large. By contrast, there is no magnetic force on an ion moving exactly parallel to magnetic field lines, so its motion is unimpeded in that direction. From these two facts we see immediately why ions and electrons more or less "stick" to field lines. An ion whose motion is only perpendicular to the field sticks because it simply executes a circular orbit. An ion moving only parallel to the field sticks because there is no force on it, so if its motion is initially parallel to the field it remains so—unless of course the line bends, as it would in a tokamak. But as the line bends, the ion, trying to continue on a straight path, begins to acquire motion perpendicular to the magnetic field line. This perpendicular component of the motion causes the ion to encircle the field line while its forward travel continues its motion along the bending line. Thus, somewhat like a car on a properly banked highway, the ion is self-guided along its "magnetic highway" by its inability to stray very far from the field line, any "straying" being instead converted to motion encircling the field line. The net result is that ion and electron orbits form tight spiral paths following the field lines. Of course, the reader is cautioned to remember that field lines are a useful fiction describing the direction of the magnetic field. The particles do execute circles as they move forward, but they encircle nothing at all.

We are now ready to begin understanding the Vlasov theory, first by applying it to calculate the equilibrium balance of forces in a tokamak. We assume that there is "charge neutrality"—that is, an equal number of ions and electrons, so that on average there is no electrostatic field due to ion and electron charges. For now, we also neglect collisions, or close encounters between particles, and consider only the magnetic field and the ion and electron motion in the field. If we knew the equilibrium field exactly, we could meticulously apply the magnetic force law to calculate one by one the orbital motion of every ion and electron in the plasma. Each orbiting particle would execute a path spiraling along the field lines, but the particles' paths would have different diameters, a different pitch of the spiral, and different directions of flow. Then we would add up all of the currents due to each of these orbits and recalculate the magnetic field, taking into account all of these currents. If we had made a perfect guess about the field we assumed in order to calculate the orbits in the first place, this recalculation would give back exactly the field we started with, and we would be done. If our guess was not quite right, we would have to adjust the field a bit and try again until we got it right. This process of repeated calculation, adjusting the field to produce the orbits that, together with the coils, produce this field, corresponds to finding the force balance in MHD theory.

As we shall see presently, it is not actually necessary to calculate the orbits one at a time. However, this graphic description of the Vlasov theory in terms of the orbits of individual particles already enables us to see how equilibrium properties such as diamagnetism and perfect conductivity emerge from the Vlasov picture.

In the Vlasov theory, perfect conductivity is simply the free flow of charged particles parallel to field lines, while diamagnetism is their circling motion around the lines. As we have seen, both follow directly from the nature of the magnetic force.

First consider perfect conductivity. Since the magnetic force parallel to the field is zero, those electrons and ions whose velocity is directed mainly parallel to the field must be able to flow round and round the torus unimpeded, and it is these so-called passing particles that carry the current. As we noted above, for the net result of particle motion to produce a current, there must be a degree of order in which, on average, electrons are flowing relative to the ions (since otherwise the randomly directed currents of differently charged particles would cancel each other). However, once a current-carrying distribution of equilibrium orbits is established, the current would simply continue to circulate, there being nothing to stop it. This is what we mean by perfect conductivity. Of course, the collisions that are not yet included in this theory or in "ideal" MHD theory would indeed offer resistance to the current, as already mentioned. Then the conductivity would not be perfect, and a voltage source (for example, the transformer in the tokamak shown in fig. 1) would be required to maintain the current.

Now consider diamagnetism. In chapter 5, we found that it is the diamagnetic nature of a plasma that allows a magnetic field to confine plasma pressure, and that this magnetic confinement is possible because a plasma in the presence of a magnetic field automatically produces a current—the diamagnetic current—that is proportional to the drop in pressure from the hot interior to the edge. The diamagnetic current can be traced to the diamagnetism of each circling ion and electron comprising the plasma. Each moving charge is effectively a circular current. Since, as we learned, the circling ions and electrons spin in opposite directions, the oppositely spinning positive ions and negative electrons actually carry current in the same direction (both clockwise or both counterclockwise), the direction of the current being determined by the direction of the field that caused the spin. Thus, even though their charges cancel, ion and electron diamagnetic currents add. This is a significant point.

Another significant point relates the diamagnetism of individual ions and electrons to the macroscopic diamagnetic current, proportional to the pressure drop, that entered into the MHD force balance. Just as we found in the macroscopic MHD picture, in the microscopic Vlasov picture as well there would be no net diamagnetic current without a pressure drop. Even though each spinning electron and ion does produce a circular current, if the particles were uniformly distributed in space with the same distribution of speeds everywhere, all of these currents would cancel owing to random overlap of the orbits. But if the particles are denser and hotter in the interior (that is, if there was a pressure drop), orbital overlap would be incomplete and a net diamagnetic current would flow. Adding up individual current contributions, we would find that the net density of diamagnetic currents (the current per unit area) at any point in the plasma is just the pressure gradient (the drop of pressure with distance) divided by the field strength. As long as the orbital diameter is small compared to the plasma dimensions, as it is in tokamaks, this gives the very same formula that we used to represent the diamagnetism of the plasma "fluid" in MHD theory.

Becoming a Certified Plasma Accountant

Useful as such mental images may be, actually calculating plasma equilibrium and fluctuations by adding up the individual effects of the billions upon billions of ions and electrons in the plasma is an impossible task, and would provide too much information in any case. What we really need, in order to consider the collective phenomena that cause the entropy to increase in a plasma, are the average electric and magnetic fields of groups of charges acting cooperatively, because each charge in the group experiences more or less the same Lorentz force due to these average fields. But what kind of average fields should we consider, and what is the best size of group to consider?

It was Vlasov's inspired idea that we do not need to know the size of the group if the group size is small compared to the spatial scale of the collective phenomena. That is, suppose we choose a group size and find that the plasma does form charged clouds but that these clouds consist of many groups moving together. Then we might just as well have picked larger groups, of the size of the clouds, in the first place. But it would do no harm to pick even smaller groups, and this is what Vlasov did, keeping only as much information as he needed about the groups. The same idea has worked in other fields of physics involving the motion of many particles.

To illustrate, suppose I have the job of surveying traffic at a busy freeway location. As a first cut, I might just count cars irrespective of color, manufacturer, and so on. A little bored, I might daydream about the cars as a blurry stream of "fluid" with a density of so many cars per mile of highway at this location. I could measure this density by using a wide-angle lens to take snapshots of a mile of highway at a time, and counting the cars on each picture. If traffic were steady (a state of equilibrium), I would count the same number in each picture, and if I narrowed the camera view to one-tenth of a mile, I would find one-tenth as many, which is, however, the same number per mile. In other words, in a steady flow I could get the correct "fluid" density per mile by sampling in groups a mile long, a tenth of a mile long, or even smaller (but not smaller than a typical spacing between cars). It is just such situations in which a "fluid" description is useful. The MHD theory is such a fluid theory, as is the Vlasov theory, but the latter provides more information.

To continue the analogy, suppose that one day I begin to notice that there are always more cars in lanes 2 and 3 than in lanes 1 and 4, so I begin to record "car density" as cars per mile per lane (two dimensions). Impressed with my work, the boss provides me with one of those radar gadgets that measure speed, so that I can count how many cars are traveling at about fifty miles per hour, how many at about sixty miles per hour, and so on, and I begin to report "car density" per mile per lane per increment of speed (three dimensions). I can expand my survey to as many "dimensions" as are needed to provide more information.

In the same way, the "density" of MHD theory describes the plasma merely as so many ions per cubic meter, without regard to their velocities, while in Vlasov theory "density" means how many, how fast, and in what direction, and density is tabulated separately for ions and electrons. Since we need three dimensions to describe space and three more to describe velocity (velocity being speed and direction), the MHD theory is three-dimensional and the Vlasov theory, carrying much more information, is six-dimensional. To make up for its lack of information about velocities, MHD theory approximates velocity information by a few numbers such as temperature (which is related to the average kinetic energy of the particles) and current (which is related to the average speed of the electrons relative to the ions). All of this information and more is contained in the Vlasov six-dimensional density, which we call a "distribution function" to distinguish it from ordinary density.

By adopting a fluid description, the Vlasov theory sidesteps the need to calculate individual orbits. To see how this works, let us again return to the traffic analogy and consider two adjacent segments of highway one mile in length. If traffic flows from left to right, I can calculate the outflow from the first, leftmost segment if I know the density and speed of cars in this segment. For example, if the first segment contains 10 cars per mile moving at fifty miles per hour, the outflow is 500 cars per hour, which is also the inflow to the second segment. If the density and speed were the same in the second segment, its outflow would match the inflow, implying a constant density in this segment. Applying this test to all pairs of segments, we could conclude, if inflows and outflows match everywhere, that the traffic flow is in steady equilibrium. Or, if the inflow and outflow do not match for some segment, we can calculate the rate of change of the density in that segment. For example, if 500 cars per hour flow in but 490 cars per hour flow out, the number of cars in the segment is increasing at the rate of 10 cars per hour, and the density there is increasing at the rate of 10 cars per mile, per hour. Adapting this kind of bookkeeping to six dimensions gives us the Vlasov equation.

Going further with the traffic analogy, recall that, using my radar, I also have gathered information about the densities of cars moving at various speeds, from which I could define a density in the two dimensions of highway distance and car speed (cars per mile per increment of speed). The Vlasov theory also employs both distances and velocities as dimensions, which is important because both cars and particles can accelerate and decelerate in velocity. Then, to get a complete picture, we need to keep track of flows in the velocity dimensions as well as the dimensions of ordinary space (three of each in the full Vlasov equation).

Just as we found a flow in space to be the density multiplied by speed (rate of change of spatial position), so also a flow in velocity is the density multiplied by acceleration (rate of change of velocity). In Vlasov theory we know the acceleration exactly if we know the fields, the acceleration being simply the Lorentz force divided by the mass (force is mass times acceleration, as Newton taught us). Thus, when we know the fields we have all the information necessary to determine the outflows and inflows at a particular position and velocity by multiplying the Vlasov distribution function, or density, at that position and velocity by the appropriate quantities (the velocity itself, to find flows in spatial directions; and the Lorentz force, to find flows in velocity dimensions). In six dimensions, this requires simultaneous evaluation of the flows in all three spatial dimensions and all three velocity dimensions.

If the sums of all inflows and outflows balance, the distribution function is unchanging at that position and velocity; and if this is true everywhere, the system is in equilibrium. If the flows do not balance, the distribution is changing, which requires us to calculate the changing distribution as time passes in order to determine what fluctuation levels might develop. The Vlasov equation is the mathematical equation with which we do this: it is a systematic way of keeping books on particle flows in space and velocity. (In the language of calculus, it is a partial differential equation in space, velocity, and time.) Since changes in the distribution function may imply changing charge and current and hence changing fields, the Vlasov equation must be solved together with Maxwell's equations for the electric and magnetic fields, using charge and current densities obtained from the distribution function. The charge density at any spatial position is obtained by adding up (integrating) the distribution function at that position for all velocities (since each charge counts whether or not it is moving), while the current is obtained by multiplying the distribution function by each velocity as we add.

Though simpler than adding up individual orbits, the time-dependent Vlasov-Maxwell equations in six dimensions still represent an almost intractable problem, even for computers, or so it seemed at the dawn of magnetic fusion research in the 1950s. A joke at that time said that the difference between a plasma theorist working in the world of classical physics and a "high-energy" physicist working at the very frontier of modern science was only this: In the morning, the high-energy theorist would arrive at the office, place a blank sheet of paper on the desk, and stare out the window. The plasma theorist would take a blank sheet, write the Vlasov equation at the top, and then stare out the window.

Little by little, first working on idealized problems in one spatial and one velocity dimension, the theorists began to crack the Vlasov problem and distill its conclusions. Some of the most important work in this early period concerned the foundations of the Vlasov theory itself, and the connection of the Vlasov theory to the more tractable but approximate MHD theory, partly summarized above. As a fluid theory, the Vlasov equation is also approximate because it averages charges and current in a way that smooths out fluctuations due to close encounters between charges, such as the binary (two-at-a-time) collisions mentioned earlier in this chapter. However, a formal analysis of individual orbital effects shows that the distribution function can be represented as a sum of various quantities, with the largest and most important being the fluid average obtained from the Vlasov equation. The effects of binary collisions are the second largest, the effects of three-particle collisions the third largest, and so on, but for fusion research it is sufficient to consider only the Vlasov equation and the binary collisions. The reason has mainly to do with the tendency of electrons to gather around individual ions (and shy away from other individual electrons) so as to almost neutralize the ion's charge as seen from a specific short distance away. (The distance is called the Debye length, and it is much smaller than orbit dimensions at high densities.) Thus the charge and current densities appear to be smooth, as assumed by Vlasov, over dimensions greater than the Debye length. Binary close encounters do produce electric field fluctuations over dimensions smaller than the Debye length, and these fluctuations are present even in a stable plasma. They can be easily accounted for by known formulas that can be added to the Vlasov equation when needed. Thus the Vlasov-Maxwell system of equations, together with collision terms, is now accepted as the complete first-principles theory for magnetic fusion research.

One of the first experimental confirmations of the Vlasov theory was a demonstration of the phenomenon of Landau damping, first described by L. D. Landau in 1946. Landau damping is ubiquitous in plasma physics and is the classic example of behavior predicted by the Vlasov equation but missing from MHD theory (or any of the other approximate theories from other branches of plasma physics). Landau's theoretical prediction of the phenomenon applied to an idealized plasma of one spatial dimension and uniform density, which was approximated in the confirming experiment by a low-temperature plasma column as free as possible from collective fluctuations. The point of the theory was the prediction that, in a stable plasma, any fluctuations deliberately created by disturbing the plasma should damp off at a rapid rate that Landau could calculate from Vlasov's equation. The phenomenon depended on a resonance between electrostatic waves created by the disturbance, and electrons in the plasma with speeds near the propagation velocity of the wave. It was this selective, resonant behavior that could only be captured by a theory that, like Vlasov's, kept track of particles with regard to both position and velocity. In 1985 John Malmberg, who had collaborated on the experiment, was awarded the prestigious Maxwell Prize by the American Physical Society for this and other pioneering work in plasma physics.

Actually, in the absence of binary collisions, Landau damping may be only temporary, since the disappearance of the electric wave coincides with the acceleration of electrons in specific ways that, unlike a truly random process, retain information about the wave. Later, in another classic experiment, Malmberg collaborated in the demonstration of the plasma "echo," in which the electric wave first disappears completely and then returns, as a group of travelers might conspire to congregate in San Francisco, disperse in travel, and rejoin each other from flights arriving simultaneously in New York.

Summing Up

My purpose in discussing the theoretical foundations of magnetic fusion research in chapters 5 and 6 has been to explain the underpinnings in sufficient detail to convince the reader that fusion science, however complicated in practice, rests on credible bedrock. And the basic facts of orbital motion that we will need are, after all, not so complicated.

Before proceeding with the practical matter of applying the free energy principle to estimate heat leakage in tokamaks, it is interesting once more to reflect upon the origins of fusion science from a historical perspective. First there was Faraday, who, in trying to understand Oersted's discovery that an electric current produces a magnetic field, discovered in turn his law of induction, which launched the age of electricity. Then came Maxwell, who, in reconciling Faraday's discovery with Ampère's law, created the classical theory of light that presaged radio, launched the age of communications, and posed the puzzle that led Einstein to discover the relativity of time, and the mass-energy equivalence that launched the age of nuclear energy.

Finally, coming full circle, scientists today are striving to use the magnetic force that started it all to harness nuclear fusion energy as a replacement for the fossil fuels that now power Faraday's dynamos. It is to be hoped that they can reach this goal in time to help forestall a modern "heat death" in the form of global warming—if only the thermodynamic heat death of the "second law" does not destroy the plasma.

7 : Law and Order

H aving delved at length into the complexities of plasma theory in the preceding chapter, let us start this one with an early admonition from Alvin Weinberg, then director of the Oak Ridge National Laboratory. Fusion scientists should always strive for "the *dullest* of plasmas," Weinberg said, meaning plasmas as nonturbulent as we can make them. The free energy principle, which is the main subject of this chapter, is a prescription for doing just that, though the task of learning how to apply the principle is far from complete.

Weinberg made his statement in 1966, in a letter quoted in Joan Bromberg's history of fusion, about a year after I had left Oak Ridge to work at General Atomics. Many of us, bewildered as one instability after another emerged from our studies of the Vlasov equation, would have agreed with him long before. Turning the tables on stability analysis has long held a special fascination for me, and many others. Instead of guessing the equilibrium state and then using the Vlasov theory to analyze the stability of this state, we would rather use the theory, if we could, to search for the most stable equilibria in a more systematic way.

Too Much Perfection

I came to this point of view as the outgrowth of my first assignment at Oak Ridge, which was to study the motion of individual ions in a mirror machine, in order to understand the experiments then being undertaken. Another colleague was already working on this problem, using the new Oak Ridge computer called the Oracle, when I arrived for work in the fall of 1957. The device we were studying was the simple magnetic mirror system mentioned in chapter 5, the one without Ioffe bars. Computers were very slow in those days, and the glitzy computer graphics now seen everywhere in video parlors did not yet exist. Our "graphics," devised by my colleague, consisted of a checkerboard and a box of metal washers. Each night the slow computer would follow the motion of an ion once across the device, in reality a trip taking only a millionth of a second for a real ion. The computer provided numbers—coordinates similar to longitude and latitude—that located the ion on the checkerboard, which was used to represent a cross section of the space inside the device. My colleague would record these numbers, placing a washer on the corresponding square on the checkerboard. Night after night, the stacks of washers grew, slowly building up a pattern indicating the path of the ion's motion. Because the coils in the simple magnetic mirror are circles, the magnetic field is symmetric around an axis joining the two coils. It was suggested to me that because of the symmetry, the ion's rotational motion about the symmetry axis would obey a conservation law, and that I should make use of this fact in trying to understand the computer results.

This was to be my first working experience with the power of conservation laws in simplifying theoretical calculations, a persisting fascination to which I have already confessed in connection with Taylor's beautiful theory of the self-organization of the magnetic field in the Zeta toroidal pinch device. It was the conservation of energy that we used in the energy principle, in that case applied to the entire plasma system. Because the total energy of the plasma is conserved, or constant in time, we could conclude with confidence that a particular plasma configuration was stable, unable to distort itself, if such distortions required additional energy. The motion of a single ion in the mirror machine also obeys conservation laws. Its energy of motion is conserved, or constant, no matter where it goes, and its rotational motion around the symmetry axis is separately conserved in a certain sense, captured in the conservation of a quantity called angular momentum that tells us the tradeoff between the speed of rotation and the distance from the axis as the ion moves against the magnetic force. Then, just as by using the energy principle we could obtain a great deal of information without actually calculating the plasma motion in detail, for the ion in a mirror machine I found that by using the conservation of energy and angular momentum I could easily calculate by hand, in an hour or less, a boundary on the checkerboard within which the ion must be located as it moved. No washers should be outside the boundary. Moreover, I could guess that in its random wanderings the ion might have an equal chance to occupy any position within this boundary, so we should expect our accumulating stacks of washers to fill the boundary, each stack being about the same height.

The agreement between this prediction and the computer result, as the stacks of washers grew just where I had said they should, was my first small success in understanding plasma behavior, an experience that only made me want to learn more. After this I spent many hours in the library pouring over old texts on Newton's classical laws of motion, such as the work of the

French mathematician Jules Henri Poincaré that I found in books on planetary motion. I applied every technique I could find in these books to our problem of ion motion in a magnetic field, mostly to no avail, wasting reams of paper.

As I would learn later, others were doing a better job, using so-called adiabatic (slowly changing) invariants, which are useful quantities that are not conserved exactly but are nonetheless usually conserved with sufficient accuracy to provide a good approximate description of the motion of particles in fusion machines. The very energetic ions then of interest at Oak Ridge were the exception, so that, though I had learned about adiabatic invariants in 1957, my attention was elsewhere. Moreover, at that time there was a great deal of confusion as to whether these not-quite-constant "invariants" could be trusted for the long periods during which ions must remain confined in magnetic fusion devices. Work on these questions in plasma physics contributed significantly to the formulation of what is now called "chaos theory," which has applications in many fields of science from quantum mechanics to the weather. Despite my fascination with such things, I played no part in the development of adiabatic invariants. I had a great opportunity to do so, but missed it.

My missed opportunity came about in the course of our work on the ion motion in a magnetic mirror device, when our computer runs began to show some peculiarities. Occasionally we began to find orbits that did not wander randomly, as I had predicted from the overall constancy of angular momentum. Instead, they traced a narrow path, hewing closely to a single line bouncing between two points on the checkerboard. I was perplexed.

About this time, we at Oak Ridge were visited by a young German, newly arrived at the Courant Institute at New York University. I described my problem and he told me the answer. The visitor was Jürgen Moser, later to become director of the Courant Institute, and well known for his pioneering role in the development of chaos theory. Our peculiar orbits, he said, resembled those he had studied in trying to explain peculiarities in the orbits of asteroids. As we would now say, these peculiar orbits did conserve a particular adiabatic invariant. More important, Moser had been able to show that, if they did, they would do so more or less forever—that is, adiabatic invariants could indeed be trusted. And he could predict sharp distinctions between orbits that did conserve the invariant and those that did not, the latter tending to wander "chaotically," as we had usually observed. Finally,

Moser's conclusions did not require that the machine be symmetric in any sense.

I think I must also have discussed with Moser the relevance of his work to similar problems in calculating whether or not magnetic field lines escape from nonsymmetric magnetic devices such as stellarators (or tokamaks, which are ideally symmetric but are actually never perfectly so). In any case, years later I was reminded by a colleague that, in bringing the news of Moser's work, I had been the first to alert Princeton theorists to the occurrence of stable "magnetic islands" in toroidal devices—regions in which field lines wander only a little if asymmetries are not too great. In plasma physics, the power of Moser's theorem and the subsequent work on chaos was to give assurance that absolute perfection, such as the perfect symmetry I had assumed in calculating ion orbits in a mirror device, is not required for the achievement of closed magnetic flux surfaces in a tokamak or the long-term confinement of individual ions and electrons.

Free Energy

For a theorist, missing the chance to get in on an exciting development such as chaos theory is like missing a chance to buy Microsoft stock at ten cents a share. However, the study of particle orbits did soon lead me to think about free energy. What I had learned, which was already well known to others, was that I could construct equilibrium states without doing all the hard work of solving the Vlasov equation explicitly, just by using the conservation of particle energy, angular momentum, and so on, which physicists refer to, collectively, as the constants of motion for individual particles. Using the constants of motion ensured that my answers would automatically satisfy Vlasov's equation. In principle, this allowed me to construct, on paper, plasma confinement devices of any shape I chose, which was a big help in searching for the most stable devices. Moreover, with a great deal of help from Alston S. Householder, then head of the mathematics division at Oak Ridge, I also learned ways to figure out whether these states were stable without solving the Vlasov equation, using methods going back to the pioneering work on stability by the Russian engineer A. M. Lyapunov and to the work of many others who had tried their hand at stability problems over the years (even Maxwell, who worked on the mechanical "governors" used to stabilize old-fashioned steam engines). I got a lot of help, also, from Clifford Gardner and others at the Courant Institute, in New York, where my family and I spent a pleasant two months in the winter of 1962. By the time we left New York, I was calculating free energy — though, being largely ignorant of thermodynamics at the time, I did not know this until a Spanish physicist visiting the University of Virginia explained to me what I was doing when I gave a lecture in Charlottesville on our way back to Tennessee.

"Free energy," in the sense in which I'm using the term, is the source of turbulence in a plasma. In the previous chapter, we learned that it is turbulence, or fluctuations, in electric and magnetic fields that causes the entropy to increase and heat to leak out of a confined plasma, as predicted by the second law of thermodynamics. We found that fluctuations occur spontaneously owing to close encounters between pairs of charged particles, and that even larger fluctuations can be caused by collective processes involving many particles simultaneously.

Fluctuations in electric and magnetic fields add to their energy. The energy in the field fluctuations associated with binary collisions was provided in the course of ionizing neutral atoms to produce the plasma, each fluctuation being the field of an individual charge in the near neighborhood of the charge, out to a tiny distance called the Debye length, as mentioned in chapter 6. Referring back to the traffic analogy in that chapter, averaging over the Debye length in the plasma is like averaging over a length many times the separation between cars in flowing traffic. Averaged over such a length, the traffic flow (or the plasma) appears as a continuous "fluid" devoid of the actual "grainy" structure due to individual cars (or particles).

The collective fluctuations in a plasma are local variations in the smooth-averaged electric and magnetic fields produced by the approximately fluid-like plasma. As noted above, these collective field fluctuations also contain energy. Hence, for collective fluctuations to exist there must be some source of energy to allow the fluctuations to develop. The free energy is that portion of the total plasma energy available to drive collective fluctuations. The less free energy there is, the weaker the fluctuations and the slower the leakage of particles and energy due to these fluctuations. Thus, to the extent possible, we should deliberately design magnetic fusion reactors so as to minimize the free energy; this is what we mean by the free energy principle. If the collective free energy could be reduced to zero, the only processes causing entropy to increase and heat to leak would be the unavoidable binary collisions. This would be the only surviving penalty exacted by the second law of thermodynamics as a price for maintaining the minimal state of order required to confine a plasma in a magnetic field. As we shall see, zero free energy is too much to ask, and it is not required for the achievement of practical fusion power.

Loosely speaking, the free energy is the energy available to do work. We encountered this idea in our earlier discussion of the MHD energy principle (see chapter 5). Applying the energy principle amounted to calculating the work or energy required to create distortions, or fluctuations, around the equilibrium state. If the value we calculated for the net work required to achieve a given distortion was positive, there was no way in which the internal energy of the plasma could cause such a distortion; and the system would be stable against small external disturbances. But if the value we calculated for the required work was negative, a portion of the internal plasma energy would be available to allow this type of fluctuation to develop. This "negative work" is the free energy. Equivalently, we could say that the distorted state was one of lower energy compared to the equilibrium state, the difference being the free energy.

The concept of free energy, which has applications in many fields of science, stems originally from the study of steam engines. When steam expands in a steam engine it pushes on a piston, thereby doing work. An automobile engine performs the same way, steam being replaced by the hot gases produced by burning gasoline. In both cases, usually much less than half of the heat energy is actually converted to work in moving the piston and whatever is attached to it, the rest being left behind as the wasted heat content of the exhaust gases. The work done is the free energy.

Free energy is closely related to entropy. Mathematically, the free energy is the change in a quantity called the Helmholtz function, defined as the difference between the plasma energy and a quantity obtained by multiplying the plasma entropy by the temperature. If I knew the correct mathematical expression for the entropy, I could calculate the free energy without having to calculate the detailed motion of the system, the same procedure followed in the MHD energy principle. And, as we shall see, given the free energy I could estimate the energy confinement time.

Educated Guesses

Although the MHD free energy is known exactly in the form of the energy principle, we unfortunately do not yet know how to calculate the entropy and free energy for the complete Vlasov problem, taking into account complicated processes such as Landau damping that are included in the Vlasov theory but omitted from MHD theory. However, using instead the known formula for the entropy for ordinary gases, we can make estimates of the free energy that help us identify the sources of plasma free energy. Moreover, these estimates are guaranteed to exceed the actual free energy, since what is missing are the constraints, not present in an ordinary gas, that couple particle motion to changes in the field energy. It was the complete coupling of particles and fields, represented as the "sticking" of plasma to field lines, that constrained the free energy in the MHD energy principle.

Setting aside, for the moment, the alpha particles produced by fusion reactions, there are in a tokamak just two potential sources of free energy of any consequence. These are the pressure or thermal energy of the DT plasma fuel, and the large current parallel to the magnetic field. Of these, it is the pressure-driven free energy that is the more fundamental, the confinement of pressure being the very essence of what we mean by plasma confinement and therefore a potential source of free energy in any scheme for magnetic confinement fusion. So, let us first discuss pressure, postponing our discussion of the special tokamak feature of the parallel current until later in the chapter.

The free energy due to pressure in a plasma is similar to the free energy of expanding steam. An electric field fluctuation, which absorbs energy as it grows, plays the role of the piston in a steam engine. The plasma supplies this energy by cooling as it expands, just as steam does, and for this reason I will refer to free energy due to pressure as "expansion free energy." It was Marshall Rosenbluth, now the U.S. senior scientist in the ITER project, who explained this to me, after I had calculated the result but was puzzled about what it meant. As the reader can see, I have always relied a great deal on my friends.

In a tokamak, the plasma pressure is not uniform, so that a graph of the pressure distribution versus minor radius resembles a sloping hillside, highest on the innermost flux surfaces and decreasing gradually to near zero at the edge of the plasma. Any expansion of the pressure releasing free energy usually occurs locally over a radial dimension that is small compared to the plasma minor radius. Afterwards the pressure distribution is roughly constant over this small dimension, flattened out, like a small landslide on the side of a hill. Such eruptions might occur anywhere on the pressure slope, subside, occur again, and so on, resulting in a turbulent state of electric field fluctuations driven by the free energy released by each local expansion event. Though different in detail, this process resembles the charging-up of clouds in a thunderstorm. After one of these small eruptions, the strip of plasma lying along the path of the little "landslide" is all charged up, either positively or negatively, like a thundercloud. It is this temporary charging-up of some portion of the plasma, thereby producing an electrostatic field around the charge, that we refer to as a fluctuation in the electric field.

As noted above, the fluctuations can be estimated from the approximate expansion free energy based on the entropy of an ordinary gas, and given the free energy one can estimate the energy confinement time. I did this in a paper published in 1965. If such an estimate had been done for the Princeton TFTR, which was still far in the future, it would have predicted for that device an energy confinement time of around one-tenth of a second. Though this is very close to the actual experimental result obtained in TFTR in 1993, I could not have forecast this with any confidence in 1965 because of the many uncertainties about the guesses I had needed to make. With this in mind, let us review my old estimate, still valid in principle, to see what we knew then and what we know now.

We should think about the calculation in three steps. The first step is to ask ourselves how the plasma can become "unstuck" from the flux surfaces. As we have seen, it is in the very nature of the motion of ions and electrons in a magnetic field that they tend to spiral around paths corresponding to the imaginary field lines that we use to visualize a magnetic field. This is what we mean by being "stuck" to the field lines. Since in a tokamak the field lines form a wonderful architecture of nested, closed flux surfaces that allow no escape, plasma stuck to the lines should not be able to escape the magnetic confinement, and its energy should be confined for as long as we keep the machine running. What is wrong with this picture?

The answer lies in the electric field fluctuations, the charging-up of the plasma after the small "landslides" discussed above. In the presence of these fluctuations, the spiraling orbit slowly drifts sideways from one flux surface to the next, and it is this drifting that allows the hot particles to escape. Turbulence of this kind is called "drift waves." Drift waves violate the MHD approximations, since the drifting plasma does detach itself from the magnetic field structure, for reasons having to do with the fact that the electrical conductivity along the line is not really perfect. However, only a short section along a field line can detach in this way.

The reason for the drift is as follows: In the absence of an electric field, an ion (or electron) in a uniform magnetic field spins on an exactly circular path with a radius such that the centrifugal force of spinning just balances the magnetic force. The greater the speed, the greater the centrifugal force and the larger the radius of the circle. We call this the gyroradius. When an electric field is present, the sideways drift motion occurs because the ion is accelerated (giving a larger gyroradius) when it is moving in the direction of the electric field, and is decelerated (giving a smaller gyroradius) when it spins around so that it is headed in the opposite direction. Due to the growing and shrinking of the gyroradius the center about which the ion is turning continually changes, resulting in a sideways drift of the center of the circular orbit. The direction of the drift is perpendicular to the direction of the electric field (as it turns out, the direction is the same for ions and electrons). So the drifting between flux surfaces that causes heat to escape requires "landslides" that produce an electric field with components parallel to the flux surfaces.

By the way, the nonuniformity of the magnetic field in a tokamak also causes the gyroradius to grow and shrink as the particle encounters a weaker or stronger field as it spins, but this drift is mostly parallel to the flux surface and therefore does little harm. There is also a component of this drift motion perpendicular to the flux surface which causes the orbit to move radially, drifting back and forth between its flux surfaces. These excursions away from the flux surface will be important in our discussion of the neoclassical transport theory later on. The excursions are greater the weaker the plasma current; with zero current all particles would escape, which is another reason that the tokamak must have a plasma current.

Continuing our story of expansion free energy: If a single expansion-driven "landslide" extended all across the plasma radius, the electric field produced by the landslide could cause particles to drift steadily outward and escape rapidly. This brings us to the second step in figuring out how to estimate the energy confinement time from the free energy. This step depends on the fact that, in a plasma that is stable according to the MHD energy principle, we usually should not expect "landslides" as large as this. In other words, in a single landslide the electric field should cause particles to drift only over a short distance. Another landslide nearby could continue pushing the particles outward, or if the new landslide happened to charge up the plasma the opposite way (say, a negative charge if the previous event had charged it positively), then the particles would drift back to their original positions, and so on, leaving to chance whether or not the net result of this random drifting in and out would actually cause the plasma to escape.

This kind of "random walk" or diffusive process is well known in physics and chemistry and leads to a simple formula for the time required for diffusion to occur. Using this formula and some simplifying assumptions, I could easily estimate the energy confinement time as the time it would take for an ion to escape in free flight (as if there were no magnetic field to stop it) divided by a small number obtained by dividing the free energy by the pressure. Since an unimpeded ion could escape very rapidly—in a millionth of a second—my old estimate of one-tenth of a second would require that the free energy be very small in comparison with the pressure, one part in a hundred thousand.

Based on a guess described below, my calculations in 1965 did predict a free energy this small, but with great uncertainty. As I remarked earlier, though we do not know how to calculate the entropy for a plasma, we can nonetheless determine an upper limit on the free energy using the well-known calculation for entropy of an ordinary gas. For the expansion free energy giving rise to the little "landslides" discussed above, the ratio of the free energy to the pressure, required in my calculation, is just the square of a number obtained by dividing the "landslide" width by the plasma radius. It is the width of the landslides, sometimes called the "mixing length" in the plasma literature, that is so uncertain. This width is an indicator of how well the plasma is "stuck" to the flux surfaces. Over dimensions larger than the mixing length, the plasma is effectively "stuck," and MHD theory applies. Over smaller dimensions, treating the plasma as a free gas is not a bad approximation. My guess, in 1965, was that the mixing length would be on the order of the ion gyroradius.

Because the mixing length was so uncertain in 1965, I soon gave up thinking about the free energy principle and turned to other things. Meanwhile, Boris Kadomtsev at the Kurchatov Institute proposed a different "mixing length" estimate of the energy confinement time, one that, for drift waves, gave results similar to mine but provided a definite prescription for calculating the mixing length on the basis of an approximate solution of the Vlasov equation. Since all such estimates involve the ion gyroradius, which becomes larger as the temperature increases, for given machine dimensions and a given magnetic field strength they predict a shorter energy confinement time as the temperature rises, as does the Goldston scaling law, encountered in chapter 3. This prediction was the basis for the theoretical concern, in 1973, that had in part motivated the United States to build the Princeton Large Torus experiment (the device that, when it achieved 100-million-degree temperatures in 1978, got me excited enough to write to President Carter).

Enter the Computer

As computers improved, scientists began to tackle the really tough problem of simulating tokamaks in detail on the computer. The large computer programs developed to do this follow Vlasov's philosophy but not his differential equation. Instead, they actually calculate the motion of thousands of particles chosen to be representative of what is happening on the scale of the mixing length, or less. As time progresses in the calculation, the electric and magnetic fields are continually recalculated from Maxwell's equations, using charge and current densities averaged over the particles, and assuming that each particle orbit actually calculated is representative of a "cloud" or "cell" of similar particles flowing together, along the lines of our earlier discussion of the Vlasov equation in chapter 6. This is called the "particle-in-cell" (PIC) method, pioneered by C. K. Birdsall at Berkeley and John Dawson at Princeton and, later, the University of California, Los Angeles. Today computer programs based on the PIC method are widely used in magnetic fusion, inertial fusion, and other plasma physics applications.

Several PIC computer programs for magnetic fusion have now been developed in the United States and elsewhere, the U.S. group of computer programs being referred to, boldly, as the "numerical tokamak." Most of these computer programs focus on the electrostatic drift waves driven by expansion free energy. The heat diffusion rates obtained from these computer simulations is qualitatively similar to those given by Kadomtsev's theory (and my free energy estimate if I used the typical mixing lengths obtained from the simulations). More important—though the last word is not yet in as I am writing this—there appears to be quantitative agreement with certain details in the TFTR data, which suggests to me that these computer models are close to achieving a predictive capability that may, indeed, justify calling them the "numerical tokamak." Applying these computer models to evaluate new magnetic confinement ideas such as those we will discuss in the last chapter of this book may help us to avoid subjecting those technologies to some of the lengthy experimental steps the tokamak had to undergo.

Having made a prophecy concerning the credibility of computer simulations, I had better elaborate a bit on what credibility entails. The goal is information accurate enough to be used in designing the next generation of devices, whatever that may be at the time. For example, in the early 1970s Robert Hirsch, the new director of the U.S. magnetic fusion research program, was anxious to take the boldest step possible to capitalize on the rising tide of the tokamak. What he wanted was a real DT plasma that achieved the Lawson criterion, which is still the goal of ITER. As we have seen, the Lawson criterion is a very specific, large value of the Lawson number—the number obtained by multiplying the plasma pressure by the energy confinement time. Since no one wanted a failure, and since the energy confinement time was still highly uncertain, especially at high electron temperatures, a stalemate developed until, at a fateful meeting in Washington that I attended in December 1973, Harold Furth, a senior scientist and later the director of the Princeton Plasma Physics Laboratory, offered a compromise.

Furth suggested that we consider break-even as the goal, less difficult than the Lawson criterion but still very ambitious in 1973. Then he began to sketch on the blackboard an idea he had published with John Dawson and Frederick Tenney in 1971, which made it possible to reach break-even at an electron temperature less than half that required for the Lawson criterion. This was the birth of TFTR, which has worked a little better and a little worse than Furth predicted more than twenty years ago, but which worked well enough to provide a convincing demonstration of the production of controlled fusion energy. Wisely, as it turned out, instead of exact break-even the Princeton researchers chose a specific output energy (ten megawatts for about a second), more easily predicted, as the first official goal of TFTR.

Today, more is demanded of theoretical prediction, because of the large size and cost of the next steps in the tokamak program, and the desirability of skipping steps in the competitive programs coming from behind. Nonetheless, the computers now available seem powerful enough to justify our taking seriously the predictions of the PIC simulations utilizing these computers. The amount of information needed in order to simulate drift waves with mixing lengths on the order of 1 percent of the plasma radius is much less than is required in many other scientific fields today, again in part having to do with the level of detail needed to make progress. The importance of the level of detail can be shown by reference to the large hydrodynamic weather simulation programs used to predict global warming trends. These programs might require one hundred hours on the largest computers available, just to provide one data point to represent the weather in a region the size of France. On this basis, I estimate that the same type of weather simulation, elaborated sufficiently to predict a rain shower in my back yard (a matter of some interest to me), would require a computing time equaling the age of the universe! Fortunately, we do not need information about plasmas at this level of detail in order to predict the energy confinement time with sufficient accuracy.

Free Energy Revisited

When I joined the Berkeley faculty in 1988 and began to acquire graduate students, I again became interested in the free energy principle and found myself, like Rip Van Winkle after his long sleep, wondering what had happened in the last twenty years. One important development had been a considerable advance in the use of adiabatic invariants to calculate particle motion. As this had proved especially useful in calculating drift waves, and since others had by then developed an elegantly simplified "gyrokinetic" version of the Vlasov equation taking advantage of adiabatic invariants, I suggested that one of my students recalculate the expansion free energy using the new version of the theory.

Alas, what I had hoped would be a new tool for inventing better fusion reactors only gave back my old answer, and uncovered a bad guess about spin that I had made before we had a proper gyrokinetic theory. But I was puzzled about the details. Still relying on my friends, I mentioned our finding to John Dawson, and learned that he had already discovered what was happening, using his PIC computer simulation. He straightened me out at his sixtieth birthday celebration, a lovely event on Catalina Island, arranged by his former students. At the party, one of his colleagues played a tape, taken from an answering machine, in which Dawson—having called at midnight to brainstorm some new physics idea—was unable to remember his own telephone number. But he does understand plasma physics like few people I know.

What Dawson had discovered was that, since drift waves can occur even in a uniform magnetic field, and since in a uniform field a constant angular momentum implies a constant velocity perpendicular to the field, the expansion free energy driving the drift wave can only come from velocity (or pressure) parallel to the field. My student had figured out that much, but with his computer simulation Dawson could see how this actually happens. Namely, as we learned earlier, when drift waves occur, only a short section of the pressure "hillside" suffers a "landslide," leaving a slightly higher or lower pressure at each end where the slide joins the undisturbed pressure slope. Due to these pressure differences, pressure then expands *along* the field lines, and it is this expansion parallel to the field that actually provides the free energy. Using his computer simulation, Dawson could also see, by the way, that one must be a little careful in equating the mixing length (over which chunks of plasma drift after a "landslide") to the original width of the slide. Once charged up, a drifting chunk of plasma can sometimes drift, or "convect," for a longer distance (a fact already known from earlier theory and experiment).

Whether or not there exist magnetic configurations that somehow suppress drift waves and expansion free energy remains an open question. From the outset, drift waves were feared as the ultimate threat to plasma confinement, the one collective loss mechanism that could not be avoided and that therefore determined the actual minimum free energy consistent with confinement, perhaps the "dullest" kind of plasmas we can have, to paraphrase Weinberg. To emphasize the point, the first version of the drift waves, discovered in the 1960s, was called the "universal mode" to emphasize its ubiquitous nature. For a while, I had thought that a toroidal magnetic well, analogous to the mirror machine with Ioffe bars, might stop drift waves, but if this is true it cannot be proved using the simplified free energy principle based on the entropy of an ordinary gas. That was the mistake I made many years ago, uncovered by the new work by my student.

Meanwhile, the main culprit in many tokamaks has been identified with a kind of drift wave that may not be quite so universal after all. This concerns the weak field region at the outer edge of the tokamak, identified in chapter 5 as the potential location for a pressure "blowout." From the particle point of view, this is a region where particles with relatively large spin angular momentum and small velocity parallel to the field lines could be trapped like the particles in a magnetic mirror device, unable to travel round and round the field lines. It was these trapped particles, both ions and electrons, which Kadomtsev and others found to be an especially likely source of drift wave turbulence.

It turns out that these ion drift waves are not universal and may indeed be suppressed by natural means except near the cold edge of the plasma. This is one of the subtle features of TFTR that is, perhaps, being modeled correctly by PIC simulations. My students and I, always on the lookout for the "*dullest* of plasmas," had run across the same point in applying Kadomtsev's mixing-length heat diffusion theory to the DIII-D tokamak at General Atomics. I had been calling this phenomenon an "angle of repose" (after Wallace Stegner's *Angle of Repose,* a Pulitzer prize–winning novel about mining in the West). Recently I find others also using this name. In Stegner's novel, later made into an opera, the angle of repose refers to the slope of a pile of slag or sand at which the friction holding the pile in place just balances the force of gravity that is trying to tear it down. In the same way, it turns out that in a plasma, unlike a free gas, there is a critical slope at which the expansion free energy of the ions in the tokamak can no longer be released. Then ions only drive strong turbulence where this critical slope cannot be maintained, at the edge where the hot plasma joins its cold environment. The term "angle of repose," though the perfect metaphor for the interpersonal friction in the novel, is a little misleading for plasmas, since it is zero friction, or perfect conductivity, that creates the perfect repose of a stable plasma in MHD theory. In any case, the somewhat analogous plasma effect should be especially evident in TFTR, to which the PIC calculations were directed, and also in the DIII-D, which my student studied. In both of these machines, though intense neutral-beam heating is able to create very high ion temperatures in the plasma interior, the actual slope or gradient of the measured temperature follows closely the angle of repose predicted by the theory.

There also appears to be a useful angle of repose for the expansion free energy of the alpha particles produced in the fusion reactions in TFTR and ITER (see chapter 4). This probably explains the fact that TFTR already successfully stores alpha particle energy equivalent to that in the core of the future ITER device. The existence of an alpha angle of repose is especially fortunate, since the alpha expansion energy is able to drive drift waves of a different kind (MHD waves), for which the mixing lengths are much larger than those for the electrostatic drift waves we have been discussing above. However, while the fuel ions and the alpha particles may exhibit such angles of repose, there is, as far as I know, no corresponding effect for the electrons in a tokamak.

Even in the absence of a natural angle of repose at which turbulence ceases, it may be possible to control turbulence in the plasma by external means. I have already mentioned the H-mode, in which turbulence is partially suppressed at the edge of the plasma. Still more exciting is a technique called "reversed shear," predicted theoretically and now being demonstrated in all of the world's largest tokamaks, in which turbulence is suppressed and energy confinement improved in the very core of the plasma. Engineers are already predicting great improvement in future power reactors with "reversed shear," and ways of testing the concept in ITER are being studied, though for now, to be on the safe side, ITER will continue to be designed by the old rules. Achieving "reversed shear" requires that the plasma current density be greatest midway between the core and the edge, whereas the conventional transformer-like inductive way of producing the current usually causes the current to peak up in the core. Thus, implementation of "reversed shear" awaited the development of other, noninductive methods for creating the current, to which we will return in chapter 10.

Magnetic Free Energy

While there may or may not exist an angle of repose for every source of expansion free energy, I have come to hope that such a "critical slope" may indeed exist for the other potential source of free energy in a tokamak, namely, magnetic energy attributable to the large plasma current that is crucial to the architecture of nested flux surfaces that makes confinement possible in a tokamak. This brings me back to Bryan Taylor's theory of the self-organization of the magnetic field in the Zeta toroidal pinch device, which is, as noted earlier, a continuing fascination for me.

One piece of evidence for self-organization in tokamaks goes by the name "profile consistency," a term introduced by Bruno Coppi to describe a phenomenon observed on the Alcator C tokamak at MIT. The phenomenon concerned the rapid restoration of temperature profiles following the application of rapid heating, even if the heat had to flow "uphill" from the outer cold region to the hot interior, a kind of angle of repose in the sense that the plasma quickly found its own way to a new stable state. More recently, similar phenomena have been observed at General Atomics. Soon a finger began to be pointed at current self-organization as well, and during the 1980s theoretical papers began to appear, by Allen Boozer and others, connecting self-organization of the current in tokamaks to the constancy of helicity, which Taylor had used to explain the Zeta toroidal pinch. As mentioned earlier, helicity conservation just means that the current wants to twist so as to become aligned with the field, and in so doing to redistribute itself, creating a roughly constant current density.

A downside to these phenomena is the potential for additional leakage of heat as changing currents disturb the magnetic flux surfaces. The mechanism by which current redistributes itself is called "tearing," a name indicating that the magnetic field lines no longer form smooth surfaces like those depicted in figure 1. Physically, the smooth current clumps into separate filaments, effectively forming skinny little tokamaks within the main tokamak.

These are the magnetic islands mentioned earlier. It is the interaction of these current clumps, or magnetic islands, among themselves that allows the overall redistribution and self-organization of the current into an ultimately stable state. Chaos theory tells us to expect this when islands become large enough to overlap each other. But while this process is going on, field lines wander back and forth between the magnetic surfaces. Hot plasma trying to follow these meandering field lines may intermix with colder plasma, allowing a more rapid transfer of heat by collisions between hotter and colder ions and electrons. Also, the electrical resistivity due to such collisions finally allows magnetic lines to break, or "tear," in order to form the final stable configuration, giving yet another opportunity for hot plasma to be derailed into colder zones. In 1978, A. B. Rechester and Marshall Rosenbluth published an estimate of heat leakage due to overlapping magnetic islands, a scary result that said that the islands must be really tiny if disaster is to be avoided.

When I began looking into these matters myself, after my Rip Van Winkle phase, I was pleasantly surprised to find that Boozer and others, using helicity conservation, had come upon a point that, I like to think, is the resolution of the magnetic turbulence problem, at least for islands large enough to warrant the level of concern created by the Rechester-Rosenbluth calculation. Actually, in a tokamak, because of the "stiffening" provided by the external toroidal field (see chapters 2 and 5), the current may be completely stable except at special magnetic surfaces where field lines exactly close on themselves. There the finite electrical resistance due to collisions can allow magnetic islands to form, but they remain quite small, never overlapping, if certain criteria are fulfilled in designing the current-drive system. More interesting to me, and the point of Boozer's work, is that magnetic turbulence can possibly be put to good use, with little unfavorable consequence due to heat leakage, because there is a stable magnetic angle of repose, and magnetic turbulence only grows to the level required to maintain this stable state. Moreover, since both theory and experiment show that the electrical resistance (along the field lines) is primarily due to binary collisions rather than collective effects, the destruction of the field due to electrical resistance is relatively slow, and becomes even slower as the temperature goes up and the collision rate decreases. Then, once the field is stable, it changes very slowly and the magnetic turbulence level required to maintain the stable configuration is very weak. From this, we can estimate the fluctuation level and the corresponding rate of heat leakage predicted by Rechester and Rosenbluth, with the conclusion that there is little penalty in allowing the magnetic field to organize itself around its own natural angle of repose. At least thirty years ago, Jim Phillips, one of James Tuck's colleagues at Los Alamos, had told me that magnetic turbulence would not matter at high temperatures; but now we had a theory to say so in quantitative terms. The most important point, long suspected, is that electrical resistance along magnetic field lines is generally "classical," being due only to the binary collisions and more or less immune even to the collective effects such as drift waves that spoil the MHD assumption of zero resistance, or perfect conductivity. As we have seen, it is this that should make magnetic turbulence diminish at high temperatures. The probable reason for this had already been identified in 1967 in a paper by myself and William Drummond, who went on to create the still-thriving fusion research program at the University of Texas. Namely, drift waves and the like typically only affect the few electrons resonant with the waves, while the remaining electrons are available to conduct current freely, as assumed in MHD theory.

From Free Energy to a Free Lunch

A magnetic field that creates itself! Here indeed is the stuff of invention, a point of great interest when we attempt to look into the future in the final chapter of this book. Meanwhile, I would like to conclude this chapter with a discussion of Boozer's own invention, a steady-state version of the tokamak that drives its own current.

The first suggestion that the tokamak might provide at least a part of its own current appeared in a paper by Roy Bickerton, J. W. Connor, and Bryan Taylor in 1971. That heat can generate electric current directly is well known; the phenomenon, occurring in certain metals, is known as the thermoelectric effect. Bickerton, Connor, and Taylor pointed out that a similar phenomenon can occur in a tokamak as a result of collisions between electrons and ions. They called this self-generated current the bootstrap current, by analogy with "picking oneself up by one's own bootstraps." The precise calculation of this effect is obtained from the neoclassical transport theory of Rosenbluth, Hazeltine, and Hinton (see chapter 6) and from similar theoretical work by Sagdeev, Galeev, Kadomtsev, and others in the Soviet Union.

The most elegant version of the neoclassical transport theory consists of a calculation of the rate of change of the entropy due to collisions, an approach that systematically relates macroscopic physical quantities such as heat and particle flow to the underlying microscopic processes, in this case the motions of ions and electrons in the magnetic field—a procedure of general validity in much of science, as first pointed out by Lars Onsager in 1931. In what may be the ultimate statement of the free energy principle, Ker C. Shaing and others have extended the calculation of the rate of change of entropy to include collective fluctuations such as drift waves, in addition to the binary collisions, again revealing Onsager-type relationships.

These Onsager-like calculations often disclose surprising connections among measurable quantities, and plasmas are no exception. As a fairly obvious example, it is not surprising that the leakage of energetic particles from a hot plasma implies a leakage of heat. However, the converse may not be true, since, like a chunk of hot metal, the plasma may cool off even though it is not leaking particles.

More surprising at first sight is the bootstrap current. As pointed out by Ira Bernstein and Kim Molvig in 1983, the origin of the bootstrap current lies in the nature of particle motion in a magnetic field, whereby random binary collisions actually cause those particles whose motion would add to the current to diffuse into the interior, while those whose motion opposes the current are slowly ejected. Recalling that the nonuniformity of the magnetic field causes orbits to drift back and forth between flux surfaces, we would find that it is those particles whose orbits make the biggest excursions that diffuse the fastest and contribute most to the bootstrap current.

By now the existence of the bootstrap current has apparently been demonstrated, albeit indirectly, in several experiments, most notably in the large JT-60 tokamak in Japan, in which as much as 80 percent of the current is thought to be the self-generated bootstrap current. Moreover, as if by a miracle, the bootstrap current is strongest midway between the plasma core and edge, just where it is needed to create the turbulence-suppressing reversed shear effect mentioned above. As we shall see, designers of future tokamak reactors are already making use of the bootstrap current to reduce the requirements for external means of driving the current, a point of great significance because reducing the electric power required to operate the reactor increases the net electrical output from the reactor. According to the neoclassical transport theory, some amount of external current drive would always be needed, since the theory says that the bootstrap current cannot exist at all in the very deep interior of the plasma. Armed with the new concept of self-organization of the current, Boozer set out to challenge this conclusion.

In a paper published in 1992, Boozer and his student Richard Weening presented calculations describing a tokamak that, in theory, totally drives its own current. The theoretical model does this because bootstrap current produced far out on the pressure slope is propagated inward by the helicity-conserving "tearing" mechanism discussed above. The outer portion of the profile is stabilized by the "stiffness" due to the external toroidal field, as in other tokamaks, while the collapse of the current in the interior creates the tiny overlapping magnetic islands that push the current inward until a magnetic angle of repose is achieved. Then the system sits there, happily sustained by a slow input of fuel to make up for particle leakage, and a supply of heat that would, in a real reactor, come from the fusion process itself. Moreover, the same model demonstrated quantitatively what goes wrong if the current distribution becomes unstable in the outer region "stiffened" by the toroidal field. There, where the magnetic islands are widely separated, either (in the stable case) they remain small, or (in the unstable case) they become so large that they destroy the plasma before a stable configuration can form, resembling the current "disruption" occasionally observed in tokamak experiments. While the Boozer model depends on helicity conservation, in 1994 John Dawson and William Nunan published the discovery of a bootstrap current in a PIC computer simulation that does not rely on this assumption. In their case, it is collective electrostatic turbulence rather than binary collisions that creates the current, again by selectively confining particles that produce the current and ejecting those that would oppose it. According to Dawson the bootstrap current, whether collisional or collective, is just another manifestation of Faraday's dynamo, the consequence to be expected if the conducting plasma tries to expand against the external fields. Neither Dawson's nor Boozer's theory has yet been tested in laboratory experiments.

This concludes our discussion of the theoretical foundations of magnetic fusion research. Looking back, we see a progressive building of confidence, first in the MHD and Vlasov underpinnings, then in the relating of theoretical examples and insights to experimental data, and finally in the development of the PIC computer simulations that hold the key for the truly predictive capability that has long been the goal of fusion theoretical research. As these computer programs progress toward greater and greater sophistication, producing reams and reams of data output, the traditional role of theory—providing paradigms and insights to aid in the interpretation of real experiments in the laboratory—will shift toward doing the same for pseudo-experiments on the computer. It is here that the free energy principle will come into its own, guiding computational experimenters toward those angles of repose associated with "the *dullest* of plasmas," and helping them to pose the right questions in the familiar language of thermodynamics.

8 : Seeing the Light

. . . .

n my early years at Oak Ridge, my office was a few miles away from the building where the fusion experiments were located. I looked forward to the day each week when our group meeting gave me the excuse to spend most of the day near the experimental facilities. There I usually had lunch with Julian Dunlap, a friend from high school who, after we had gone our separate ways, had taken a job as an experimentalist in the Oak Ridge fusion program at about the same time that I arrived there as a theorist. Dunlap always had a good lunch, which I would share after finishing mine. But what I looked forward to most was hearing about the latest measurements in the laboratory.

It was the experiments of Dunlap and his colleagues that had inspired my simple calculation, mentioned in chapter 7, of a boundary in space that should contain the orbits of energetic ions trapped in a magnetic mirror machine. Today experimenters can "see" what is happening inside a plasma by very sophisticated means, many involving light and other electromagnetic radiation emitted or reflected inside the plasma. But in 1958, while my theoretical colleague and I attempted to trace the boundary of ion motion in a mirror machine by computer, Dunlap and other experimentalists did this in the real machine by gradually inserting and removing a "paddle" to locate the boundary by determining the paddle position at which ions first began to intercept the paddle. In this way they could trace the entire boundary by moving the paddle from place to place. I was, of course, happy when the boundary they found resembled closely the one traced out by the computer.

Relatively crude measuring devices such as the paddle were typical in the early days. Some of these, like the "Langmuir probe," were inherited from earlier studies of gaseous discharges. Irving Langmuir and others pioneered such work at General Electric. The Langmuir probe is a tiny wire inserted into the plasma, insulated except for a small tip left exposed to collect ions and electrons striking the bare wire. Wires connect the probe to meters outside the vacuum vessel that measure the electrical current in the probe wire, and the voltage on the probe can be varied by introducing a battery or other voltage supply in the circuit.

In a landmark paper published in 1924, Langmuir and H. Mott-Smith presented a theory whereby the plasma density and electron temperature could be determined from the probe voltage and current. That a "theory" of

a measurement is needed is a subtle point, relevant to most physical measurements, and therefore deserves some discussion. The theory might be
very simple, as in the paddle experiment, which served only to locate the
plasma; but sometimes the theory is complex, and the interpretation of a
measurement may be highly dependent on the theory used to arrive at this
interpretation. We have already encountered a famous example in magnetic
fusion research in chapter 3, concerning the debate over how to interpret the
Russian measurements of electron temperature in the T-3 tokamak. It was
only when the equipment needed for the more sophisticated laser technique
was airlifted to Moscow that scientists in the West became convinced that
the T-3 results were indeed the breakthrough that they have proved to be.
But, as we shall see, even the laser technique requires interpretation; as does
the electrocardiogram your doctor may have suggested to check on your
heart, and the Dow Jones industrial average, and almost any other attempt to
reduce complicated things to a few meaningful numbers. In the end, confidence in the interpretation of scientific measurements rests upon consistency between different techniques that should, if the interpretations are correct, measure the same thing; and in the increasing detail provided by the
measurements.

Despite their simplicity, probes inserted into the plasma provided some
of the earliest tests of theory that began to build researchers' confidence in
their understanding of plasmas. For example, probes of various sorts provided the first information on the collective fluctuations in plasmas discussed at length in the previous chapter. In measurements of this type, a
fluctuation is detected as a rapid oscillation in the current flowing from the
plasma and through cables connecting the probe to recording instruments
located some distance away, outside the vacuum vessel. Like signals on a
poor telephone line, the high-frequency signals were attenuated in traveling
such distances unless the experimenters took great care to design their
probes and cables to transmit all of the information desired.

While probes continue to be useful in the smaller plasma experiments
typical of research in universities, measurements deep inside the hot interior
of large tokamaks require other techniques, which do not interfere with the
plasma as physical probes do, and which do not require the insertion of instruments that would themselves be damaged by the hot plasma. As already
noted, many of these techniques rely on electromagnetic radiation (light,
microwaves, and so on) emitted naturally from the plasma, or externally
generated microwave and laser beams pointed through the plasma. Some

techniques focus on the electrons, some on the ions. In the remainder of this chapter, we will look at an example of each.

For the last seven years, several of my students and postdoctoral associates at Berkeley have been involved in ion measurements on the DIII-D tokamak at General Atomics, in San Diego. The DIII-D is the second largest tokamak in the United States, after TFTR at Princeton. Like all major fusion experiments, the DIII-D requires a large team of scientists, engineers, and technicians to plan, execute, and interpret experiments and to maintain and operate the facility. Since I wanted my students to get this kind of experience, I had arranged our participation in DIII-D because it was an exciting opportunity not too far from Berkeley. (Students from many other universities also participate in DIII-D, and in TFTR.)

The DIII-D, now one of the most prominent tokamaks in the world, has had an interesting history. Though totally funded by the U.S. Department of Energy, it is the only surviving magnetic plasma confinement experiment in the world built and managed by a private company on its own premises. The DIII-D has also had an interesting technical history. It began in the early 1960s as a creation of scientist-inventor Tihiro Ohkawa, long-time director of the DIII-D program and a recipient of the American Physical Society's prestigious Maxwell Prize.

I first met Ohkawa in the summer of 1954 at the University of Michigan, where I was a student participant in a particle accelerator design project headed by Donald Kerst, of the University of Illinois. Incidentally, Kerst was the inventor, in 1940, of another kind of accelerator called the betatron, which, in its plasma version, is very much like a tokamak designed specifically to accelerate the "runaway" electrons that the Russian scientists were trying to avoid in the T-3 tokamak. Ohkawa had been invited to the United States from Japan when it was learned that, though still very young, he had already independently invented a new kind of accelerator similar to the one being developed by the researchers at Michigan. Later, when Kerst decided to get into fusion work, Ohkawa went with him to General Atomics, where we met again when I, too, moved there in 1965.

I always enjoyed hearing about Ohkawa's ideas and trying my hand at calculating how they would work out. When funding problems caused me and others to move on to the Lawrence Livermore National Laboratory in 1967, I tried to convince Ohkawa and his friend Masaji Yoshikawa to go with me. Though the Livermore officials did make them an offer (the laboratory's first ever to noncitizens), Ohkawa and Yoshikawa decided to stay at General

Atomics and try to rebuild the fusion program there with new sources of funding, concentrating on one of Ohkawa's ideas that eventually became the DIII-D tokamak. Later, Yoshikawa returned to Japan, where he headed the huge JT-60 tokamak project for several years before moving still higher in Japanese science and administration.

Ohkawa's original idea, called the "doublet," was a torus with a cross section in the shape of a peanut. Though we liked to joke that Ohkawa had chosen this shape in anticipation of the election of peanut-farmer Jimmy Carter as president, the real motivation (based on MHD theory) was to achieve record values of the plasma pressure relative to the magnetic energy required for its confinement, an important parameter in fusion reactor design. In time, the doublet shape was abandoned in favor of a cross section shaped like the letter D, also inspired by the evolving MHD theory. Whereas the larger TFTR project has stayed with the more conventional circular cross section, the JET facility also adopted a D shape, albeit one less extreme than that of the DIII-D. The shift to the D shape in the Doublet machine became a joint venture with Japan that helped provide operating experience for Japanese scientists in training for their own JT-60, which was still under construction. Thus, when the experiments at General Atomics did set world records, the results were presented in joint publications by American and Japanese scientists. Pressure data from DIII-D played an important role in providing experimental verification of Troyon's MHD calculations and the pressure rule, discussed in chapter 2.

The DIII-D was already thriving when, in 1988, I began looking for suitable thesis projects for my graduate students. By then, in addition to doing pressure measurements, the DIII-D team of experimentalists had become involved in measurements required to determine the energy confinement time and heat transport in detail. Though not yet as sophisticated as that of TFTR and JET, the DIII-D's instrumentation was, like that of these bigger devices, largely automated, so that much of the theoretical processing of the data was carried out instantly on computers programmed specifically for the purpose. For example, the computer could automatically produce a map of the flux surfaces and the density and electron temperature on each surface.

An exception was the DIII-D ion temperature data, which still required much hand labor and human intervention. Here, then, was an area where my students could provide needed help, and learn in the process. Though skeptical at first, the General Atomics scientists gave generously of their time to help the students. I like to think that progress by these students contributed to the eventual automation of the ion temperature measurements on DIII-D, and, as an unanticipated bonus, a quite different ion measurement that we will turn to now as an example of a sophisticated diagnostic tool foreshadowing what we may expect in ITER.

The diagnostic method used to measure ion temperature in DIII-D is called charge exchange recombination spectroscopy (CER, or sometimes CHERS, for short). The CER measurements involve detecting light emitted by ions deep inside the plasma. For his thesis topic, one of my students who had learned how to analyze CER data decided to apply what he had learned to measure not the temperature but the density of alpha particles, as part of a team that included scientists from Oak Ridge who had come to San Diego to work on DIII-D.

We encountered alpha particles, or helium nuclei, in chapter 4, alphas being the particles produced by fusion reactions in TFTR. There we were concerned with whether or not the energetic alpha particles produced by fusion would stay in the machine long enough to heat the DT fuel. My student's thesis problem addressed an opposite concern—whether, after depositing most of their energy by collisions with the fuel ions, the inert alpha particles would accumulate like ashes, to the point that the fusion "fire" could no longer burn.

Since the DIII-D is not equipped to carry out experiments with actual DT fuel, the experiments on alpha or helium "ash" in DIII-D are simulations relying on alphas produced from helium gas that happens to be introduced routinely into the deuterium plasma in the course of preparing the machine for operation. The purpose of the experiments in which my student participated was to measure the density of alphas produced from the helium gas, at each flux surface. Theorists would then utilize this information to determine the extent of accumulation of the alphas and their rate of leakage out of the system, in order to predict whether or not alpha accumulation is likely to pose problems for ITER.

The quantities actually measured in CER instrumentation are tiny flashes of light emitted by helium ions inside the plasma. The intensity of the light indicates the density of the helium ions emitting the light. The location of the source of light, and hence the location of the density of helium ions being observed, is determined rather precisely, in the following way: A helium ion in a hot plasma would lose both of its electrons by collisions, in which case it could not radiate. The "recombination" light being observed occurs only if a neutral atom collides with the helium ion. In such a collision, an electron may jump from the neutral atom to the helium ion, a process called charge exchange. Electrons that do so tend to arrive in what is called an "excited state," which means that the electron has an excess of energy. This energy is quickly radiated as light at a characteristic frequency that identifies the source of the light as an excited helium atom. This is the flash of light used to detect the presence of a helium ion in the CER measurements. Moreover, since only helium ions that have recently collided with a neutral atom can radiate, we would know where the helium ion is located if we know where the neutral atom is. Ordinarily, there are no neutral atoms available for charge exchange deep inside a hot plasma, since cold neutrals would soon become ionized. However, beams of energetic neutral atoms used to heat plasmas can indeed penetrate deeply into the plasma, and it is just such a neutral beam that is utilized to excite helium ions in the CER measurements. (We will learn more about neutral beams in chapter 9.) What we need to know now is that such a beam can be accurately pointed through the plasma. Then, if we observe helium recombination light, we know that the helium ion is located somewhere along the well-defined path of the neutral beam. Moreover, if we observe the light through a system of lenses that, like a telescope, can also be pointed accurately, we know that the helium ion is located at the intersection of the beam and a line-of-sight path through the lens system. In the DIII-D experiment the helium light emitted from helium ions along the path of two different beams was observed at numerous locations by means of thirty-two systems of lenses and mirrors that defined different lines of sight through the plasma. From this amount of data, one could obtain the helium ion (alpha particle) density at specific locations on different flux surfaces. Then, if the density was constant on a flux surface (as expected, since the ions travel fairly freely within a surface), we would know the entire helium ash distribution with fair accuracy, limited by the number of observation points.

I mention all of this detail to give the reader a feeling for the completeness of the information being gathered in magnetic fusion experiments today. Even so, this information is only as accurate as the experimenter is careful—for example, in locating the optical systems precisely in the correct position and calibrating the detection equipment that translates the intensity of light observed into numbers. Thus my student spent many hours inside the DIII-D tokamak (where he could stand erect, though he is more than six feet tall) locating and measuring known light sources, as others had done in locating and aligning the complex optical systems that defined the paths of viewing. All of this takes teamwork and the patience to wait one's turn, sometimes for weeks, as project managers try to fit numerous experiments into a crowded schedule disrupted by unforeseen emergencies.

When sophisticated measurements are made, "corrections" are sometimes needed, often prompted by mysteries discovered in the course of making the measurement — departures from the ideal concept that motivated the measurement, which require that the theory of the measurement be augmented to take them into account. In the case of my student's application of the CER technique to measure helium ion densities, the mystery was the occurrence of small differences in results depending on the direction of viewing. A suspected problem was the "plume effect," identified earlier by Raymond Fonck, then at Princeton. The basic idea is that in addition to excited helium ions created directly within view, others created along the neutral beam path can drift along field lines into the field of view. Confronted with this problem, the student decided to write an additional computer program to calculate these effects, in order to resolve the uncertainties. In other words, he had to elaborate upon the theory behind his measurement interpretation.

As an example of measurements focusing on electrons, I would like to return to the laser technique to measure electron temperature that the British team airlifted to Moscow back in 1969, which is still the workhorse of electron measurements. Incidentally, as I am writing this I am awaiting the outcome of another such electron temperature experiment, this time in a laboratory four floors below my Berkeley office. There a colleague and his student are attempting to measure an increase in electron temperature that occurs when strong microwave heating is applied to the plasma confined in a small device called a spheromak, which we shall meet again (in chapter 16) when I discuss new ideas on the horizon. One key point that could be elucidated by this experiment is whether or not this type of device, potentially simpler than a tokamak, can confine heat as well as a tokamak can. Like the Russians, we found ourselves with more than one interpretation, a dilemma that only a definitive measurement of the electron temperature can resolve. But we had no laser. Fortunately, being at a university, we were able to borrow the necessary equipment from the Department of Energy laboratory at Livermore fifty miles away—not an airlift exactly, but a "truck-lift," so to speak.

Since the laser technique for measuring electron temperature has played such an important role in magnetic fusion research, I should explain it in some detail. Plasma physicists call this a "Thomson scattering" measurement, referring to the fact that it is the deflection or scattering of the laser light by the plasma electrons that provides the information. The light is scattered by the plasma for the same reason that light can be scattered in passing through glass or other material, which also contains electrons. Thomson scattering, named for the J. J. Thomson who first discovered electrons in 1897, is "incoherent" scattering in all directions, like the scattering of light coming through a cloudy shower door.

Thomson scattering yields data on both the density and the temperature of the electrons. The measurement is similar to the CER technique described above, in that the point of observation is the intersection of the laser beam and a line of sight along which the scattered light is viewed. However, the Thomson scattering measurement is much more precise because the laser beam is a very well defined, pencil-sharp bundle of parallel light waves, and there are no confusing effects, such as the CER plume effect, inside the plasma. There are other complications, however, such as stray light reflected from the walls of the machine. To provide accurate measurements, the tiny fraction of the laser light that is scattered must be great enough compared with all extraneous signals.

Information on the electron density is provided by the ratio of the intensity of the scattered light to the laser power, a quantity that is typically no more than one part in a trillion. The temperature measurement depends on the fact that light scattered from moving electrons is actually re-radiated by the electron and suffers a change in wavelength (or frequency) due to the electron motion, like the changing pitch of a train whistle moving past you. Then, if the intensity of scattered light is measured at different wavelengths, the spread in wavelengths around the original wavelength of the laser light gives a measure of the electrons' speeds, and hence their temperature. Because of the very precise location of the laser beam, the electron density and temperature can be measured at many locations if many viewing channels are provided.

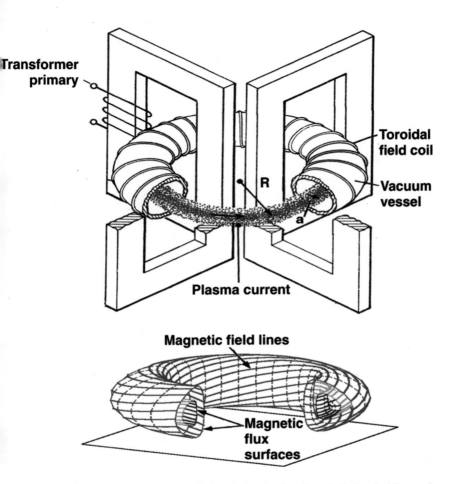

Transformer primary

Toroidal field coil

Vacuum vessel

R

a

Plasma current

Magnetic field lines

Magnetic flux surfaces

FIG. 1. *Above:* A tokamak magnetic fusion device. In the tokamak, the hot fuel forms a low-density plasma inside an evacuated vessel shaped like a torus, or doughnut. The plasma is confined by a strong magnetic field created by two sources of electric current, one flowing in "toroidal field coils" wrapped around the vessel and another induced in the plasma itself by a transformerlike primary circuit. *Below:* A representation of the magnetic field inside the toroidal vessel as imaginary lines of force forming closed surfaces nested one inside the other. The plasma, which is effectively "stuck" to the field lines, cannot easily escape. Courtesy of the Lawrence Livermore National Laboratory.

FIG. 2. The definitive demonstration of controlled fusion energy on Earth began at Princeton University on December 9, 1993, in the Tokamak Fusion Test Reactor. The photograph above was taken inside TFTR during a shutdown of the machine. When operating, TFTR has produced up to 10 million watts of fusion power for about a second, during which time the plasma at the location where the man is kneeling reached temperatures up to 300 million degrees Celsius. The tiles mounted on the centerpost in the foreground protect machinery within from heat and plasma bombardment during operation. Courtesy of Princeton Plasma Physics Laboratory.

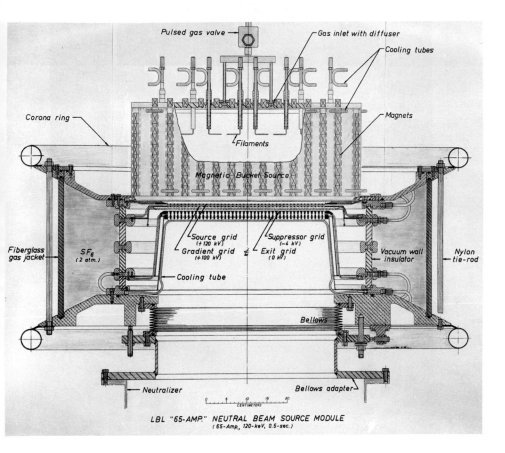

Pulsed gas valve

Gas inlet with diffuser

Cooling tubes

Corona ring

Magnets

Filaments

Magnetic Bucket Source

Source grid
(+120 kV)

Suppressor grid
(−4 kV)

Gradient grid
(+100 kV)

Exit grid
(0 kV)

Fiberglass
gas jacket

SF₆
(2 atm.)

Vacuum wall
insulator

Nylon
tie-rod

Cooling tube

Bellows

Neutralizer

Bellows adapter

CENTIMETERS

LBL "65-AMP." NEUTRAL BEAM SOURCE MODULE
(65-Amp, 120-keV, 0.5-sec.)

FIG. 3. A design drawing, in cross section, of the neutral beam injector developed by the Lawrence Berkeley Laboratory to heat the plasma in TFTR. Forty of these injectors were needed for the record-setting production of fusion energy discussed in chapter 4. Deuterium or tritium gas introduced at the top forms a low-temperature plasma whose ions are extracted and accelerated (downward in the figure) by applying 120,000 volts across the grids at the center (see chapter 9). Neutralized in flight, the fast ions become fast beams of neutral atoms directed through ports of entry into the TFTR plasma. Only a meter in height, one such unit sits comfortably on a desktop. Courtesy of Ernest Orlando Lawrence Berkeley National Laboratory, University of California.

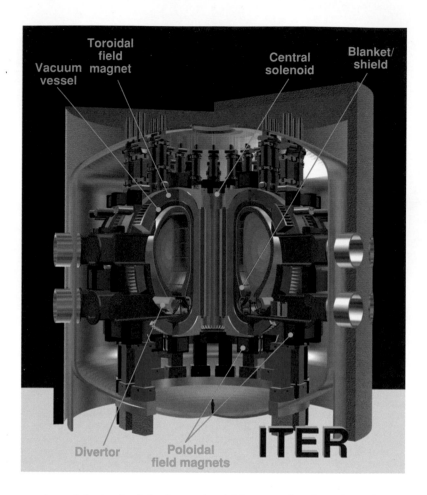

FIG. 4. An artist's sketch of the International Thermonuclear Experimental Reactor, which is being designed to be the world's first magnetic fusion reactor that will ignite and burn by itself. Producing about 1,500 megawatts of fusion power, more or less continuously, ITER would also serve as a testbed for fusion engineering development. In this design a *central solenoid* serves as an "air core" transformer to drive current in the plasma; the *toroidal field magnets* consist of twenty large coils distributed around the torus; and *poloidal field magnets* provide vertical centering and stretch the plasma into an elliptical cross section. Note especially the blanket-shield nuclear engineering components and the divertor to handle exhaust heat (discussed in chapter 10). Courtesy of the Lawrence Livermore National Laboratory.

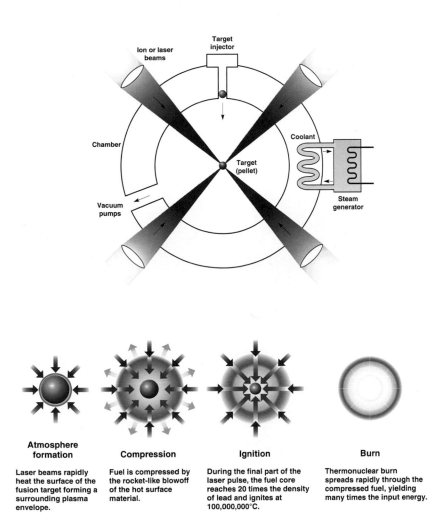

Ion or laser beams	Target injector
Chamber	Coolant
Target (pellet)	
Vacuum pumps	Steam generator

Atmosphere formation

Laser beams rapidly heat the surface of the fusion target forming a surrounding plasma envelope.

Compression

Fuel is compressed by the rocket-like blowoff of the hot surface material.

Ignition

During the final part of the laser pulse, the fuel core reaches 20 times the density of lead and ignites at 100,000,000°C.

Burn

Thermonuclear burn spreads rapidly through the compressed fuel, yielding many times the input energy.

FIG. 5. *Above:* An inertial confinement fusion reactor. Tiny fuel pellets injected several times per second are ignited by lasers or ion beams. Energy produced by these rapid "microexplosions" heats the target chamber to produce steam to make electricity. *Below:* To burn efficiently, a pellet must be made to implode upon itself, being compressed to very high density in a series of steps all of which occur in a billionth of a second. Courtesy of the Lawrence Livermore National Laboratory.

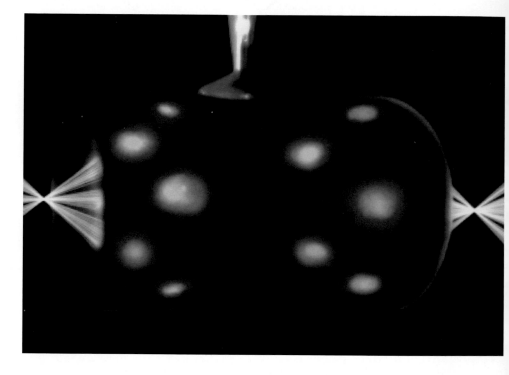

FIG. 6. An x-ray photograph, highly magnified, showing hot zones where laser beams strike the hohlraum wall in an experiment using the Nova laser at the Lawrence Livermore National Laboratory. The laser beams, ten altogether, enter the hohlraum cavity through small holes at each end. The outer shell of the hohlraum and the beam paths, actually invisible in an x-ray photograph, were added by an artist. Courtesy of the Lawrence Livermore National Laboratory.

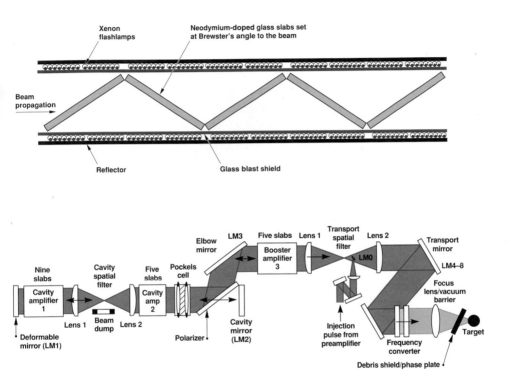

FIG. 7. *Above:* Laser beams are created by passing a weak "trigger" beam through an amplifier. In the glass lasers most commonly used in ICF experiments, energy to amplify the beam is first stored in the glass by flashlamps that "excite" neodymium atoms that retain their energy until stimulated by the passing trigger beam. *Below:* A complete beamline consists of a series of amplifiers, such as the "Beamlet" prototype beamline for the National Ignition Facility, shown here. As a cost-saving measure, in this design amplifiers are used repeatedly; the beam is passed back and forth several times before it emerges on the right, where the near-infrared light produced by neodymium is converted to the ultraviolet light required for ICF experiments (see chapter 13). Courtesy of the Lawrence Livermore National Laboratory.

The National Ignition Facility—192 Beam

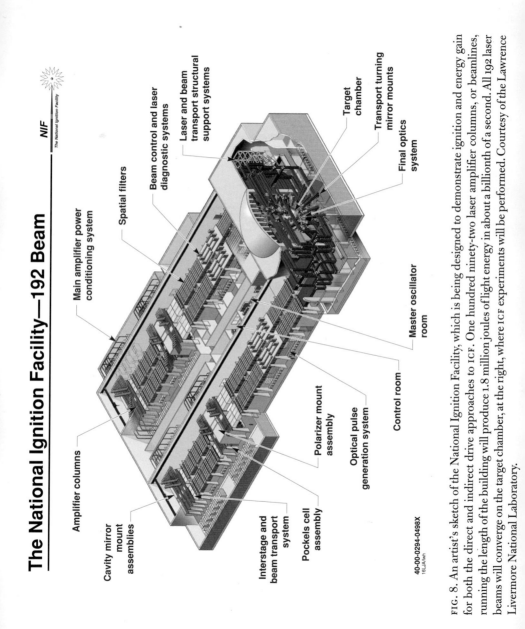

Amplifier columns

Main amplifier power conditioning system

Spatial filters

Beam control and laser diagnostic systems

Laser and beam transport structural support systems

Cavity mirror mount assemblies

Interstage and beam transport system

Pockels cell assembly

Polarizer mount assembly

Optical pulse generation system

Control room

Master oscillator room

Target chamber

Transport turning mirror mounts

Final optics system

NIF
The National Ignition Facility

40-00-0294-0498X
16LLA/lwn

FIG. 8. An artist's sketch of the National Ignition Facility, which is being designed to demonstrate ignition and energy gain for both the direct and indirect drive approaches to ICF. One hundred ninety-two laser amplifier columns, or beamlines, running the length of the building will produce 1.8 million joules of light energy in about a billionth of a second. All 192 laser beams will converge on the target chamber, at the right, where ICF experiments will be performed. Courtesy of the Lawrence Livermore National Laboratory.

9 : Engineering the Tokamak

Though magnetic fusion research at Livermore, where I worked from 1967 through 1987, was primarily concerned with mirror devices, my first decision as a manager soon got us deeply involved with the tokamak program at Princeton. The story began during the summer of 1970. My family and I were spending a month at the International Centre for Theoretical Physics near Trieste, where I was enjoying my last fling as a working theorist before assuming new duties as an associate director at Livermore, heading up its magnetic fusion research program. While at Trieste, I received an urgent message from my predecessor, who was soon to retire, asking how I thought we should absorb a reduction in our funding.

At that time the Livermore fusion program had budgetary responsibility for a small group of researchers at our sister laboratory in Berkeley, now the Ernest Orlando Lawrence Berkeley National Laboratory (referred to here as LBL). The LBL fusion group, headed by Wulf Kunkel and the late Robert Pyle, had an excellent reputation in plasma physics and atomic physics relevant to plasmas, and LBL as a whole, founded by cyclotron-inventor and Nobel laureate Ernest Lawrence, was world-famous in particle accelerator design and construction. As it happened, my colleague Richard F. Post at Livermore had been thinking about how to build a neutral-beam heating device very much more powerful than anything ever attempted. The device, called a "neutral-beam injector," is a small accelerator that creates a beam of ions that steal electrons from gas atoms and thus turn into a fast beam of neutral atoms in flight. Post was asking for only a modest voltage, as accelerators go, but in order to have a very high power output he wanted an ion beam with ten or more amperes of current, thousands of times greater than the neutral beams then being utilized in mirror research.

My decision back in Trieste was to approach Kunkel and Pyle, when I returned home, about giving up their fundamental research on atomic and plasma physics in order to concentrate their efforts on developing Post's ten-ampere neutral beam. I knew that they had the right expertise. The heart of the device is the "ion source," a small chamber into which gas is introduced and where it is ionized to form a plasma. Ionization takes place by the collision of gas atoms with electrons emitted from hot tungsten filaments, as in old-fashioned radio tubes. The ion beam is created by placing two grids of wires, one above the other, over the open side of the ion chamber. The application of a positive voltage between the two grids accelerates any plasma ion that strays between the wires, leaving electrons behind. The secret of a good ion beam is a uniform cold plasma in the ion source—very much in the realm of plasma physics and atomic physics in which the LBL group excelled.

After much soul searching, Kunkel and Pyle agreed to accept my challenge to develop powerful neutral beams. These LBL beams, and the more powerful ones that followed, made possible the growing magnetic mirror program that thrived at Livermore for the next fifteen years, and also the present twenty-megawatt neutral-beam injector system on the DIII-D tokamak at General Atomics, and the forty-megawatt system that finally produced fusion power in TFTR at Princeton in 1993. Parallel developments in Europe and Japan provided the neutral beams for JET and the JT-60, respectively.

The interest in neutral-beam injectors stems from the fact that they provide a powerful means of heating a plasma that can be produced outside the vacuum chamber and then transported into the chamber. Early on, at Oak Ridge and in Moscow, very energetic ion beams had been used to produce hot plasmas in magnetic mirror devices. However, because magnetic fields cause charged particles to spin in circles, these ion beams could only inject particles to a depth determined by the ion gyroradius in the magnetic field. The neutral-beam injector circumvents this problem by turning the ion into a neutral atom in flight. This neutral atom, unaffected by the magnetic field, continues on a straight path into the plasma, where, upon colliding with electrons and ions already present in the plasma, it becomes ionized again. The energetic neutral, once ionized, stops in its tracks, spinning around the field line where the ionizing collision occurred. Now confined, the ions accumulate to form a hot plasma; or these accumulating hot ions can serve to heat a denser plasma already present.

The assembly that constitutes a neutral-beam injector—the combined unit, including the ion source, the accelerator, and a portion of the charge exchange neutralizer—is shown in figure 3. The unit is very compact, the first ten-ampere unit being a cube about ten centimeters on each side that produced two hundred thousand watts of beam power. This figure is an actual engineering drawing, an inside view as if the device had been sawed in half vertically. Above is the ion source (labeled *magnetic bucket source*). As described earlier, gas injected at the top (labeled *gas inlet*) is quickly ionized by electrons emitted from hot tungsten filaments (labeled *filaments*), thereby forming the plasma from which ions are extracted to create the beam. At the bottom of the ion source is a grid of wires (labeled *source grid*) that ordinarily would prevent the escape of the plasma if the wires are closely spaced (as a screen keeps out flies), but when a positive voltage of 120,000 volts (120 kilovolts) is applied, any ions poking through the holes are snatched away—accelerated—to create the beam. Actually there are four accelerator grids: the source grid and the *exit grid*, with 120 kilovolts of voltage between them; an intermediary *gradient grid*; and the *suppressor grid*, with a smaller negative voltage to stop stray electrons. Together this system of grids causes ions to be accelerated along almost parallel paths at right angles to the grids, forming a uniform ion beam directed downward in the figure. Finally, each ion becomes a neutral atom as the beam passes through the neutralizer at the bottom and continues on its straight path into the plasma, meters away. To maintain a vacuum, the neutral-beam injector is mounted at one end of a long tube that leads between magnet coils through a hole in the wall of the vacuum vessel. In large installations, two or three injectors can be mounted on a single tube, or "beamline."

The first application of the LBL neutral-beam injectors was in the mirror experiment 2XII at Livermore. (The name "2XII" just means the second device [II] in a series of devices twice [2×] as big as the one before.) The first beam tested on the 2XII machine, in 1973, produced such encouraging results that a much more ambitious program was soon proposed. This new mirror experiment, 2XIIB (B for "beams"), was to have twelve neutral-beam injectors operating simultaneously, each five times as powerful as the first ten-ampere LBL beam. The new design was rectangular, essentially comprising five of the smaller beams side by side. This became the prototype for the neutral-beam injectors for TFTR and DIII-D, which differed mainly in having an acceleration voltage much higher than the 20,000 volts in 2XIIB—120,000 volts and 80,000 volts, respectively.

The 2XIIB neutral-beam injection experiment launched a resurgence in mirror research and played an important part in neutral-beam development for tokamaks. In the summer of 1975, under the leadership of Frederick Coensgen, the 2XIIB mirror device produced the world's first plasma at the 100-million-degree-Celsius ion temperatures required for fusion reactors, three years before this was achieved in the PLT tokamak at Princeton, also using neutral beams. This was an exciting event at Livermore, which established the mirror as the leading U.S. alternative to the tokamak for a decade.

I still remember vividly the telephone message from Coensgen telling me of his success, just in time for a meeting the following day of a fusion advisory committee sent to judge our progress. Coensgen's results were so recent that, when he made his presentation, the picture with the key data was streaked with rain as he ran from one building to another where the committee was meeting.

Lest this sound too easy, let me mention just one of the many difficulties whose solution in 2XIIB helped clear the way for the multiple-injector neutral-beam systems in TFTR and DIII-D. The problem had to do with "cross-talk," whereby an electronic message intended to fire a specific beam instead triggered another. This could happen because electric currents traveling along a cable between the control systems and a particular injector could radiate electromagnetic signals that would then be picked up by cables to other injectors. In a later mirror facility called the Tandem Mirror Experiment (TMX), Coensgen and his team avoided cross-talk by using fiber optics, a then-novel technology that sends messages via light pulses traveling along plastic fibers. This worked very well, except for one unfortunate incident in which someone inadvertently drove a heavy cart over the new optical cable and crushed the fibers to bits.

Meanwhile, the Oak Ridge National Laboratory had also embarked on a neutral-beam development program, initially to provide beams for its own early tokamak experiment, the Ormak. It was Oak Ridge that first accepted the challenge from Washington to propose a DT-burning plasma experiment using neutral-beam heating. Two of the bright young leaders at Oak Ridge were themselves to become leaders in the Department of Energy's Office of Fusion Energy, in Washington—John Clarke as director from 1981 to 1990, and Michael Roberts, now director of international programs.

When Princeton won the DT competition with its TFTR proposal (see chapter 7), Oak Ridge offered to provide the neutral beams for the Princeton PLT experiment and then TFTR itself, in competition with the LBL team that had developed the beams for the Livermore mirror program. Since by this time the group at LBL had grown in strength sufficiently to operate independently, my own role in this friendly competition was minimal. In any case, the Princeton scientists were well acquainted with LBL's performance and in the end chose LBL to provide the beams for TFTR.

I will mention one incident in which I was personally involved, as an example of the intertwining of science and policy. This concerned a wing of a building at Livermore that was to be used as a test facility for the neutral beams LBL was developing. Since Oak Ridge had a similar facility, the Washington managers decided that two were not needed and that ours should be closed in order to preserve some competitive capability at Oak Ridge. While I could agree that having more than one development team was healthy, this decision to close the Livermore test facility bothered me because our facility had neutron shielding that would allow testing the actual deuterium beams needed for a DT experiment, while the Oak Ridge facility did not have such shielding. After stewing a bit, I finally called Robert Hirsch, the fusion director in Washington, to point out the shielding problem. When he realized that I was calling at 4:00 A.M. my time, he knew I was serious. We kept the test facility, which was later used to test deuterium beams for TFTR.

The tokamak inherited from magnetic mirror research not only neutral beams but also other means of plasma heating and, to some extent, "superconductivity" technology for magnets in future tokamaks such as ITER. Though I cannot say that this was altogether intentional, it is also no surprise to those familiar with the highly interdependent manner in which complex technologies evolve. Generally, new technologies have been introduced by necessity, either to solve a problem or as the means of opening some whole new approach to fusion.

Consider, for example, magnet technology. When I entered the field in 1957, all magnetic fusion experiments employed magnets that contained copper conductors and relied on various techniques for disposing of or reducing the enormous heat generated by electrical resistance to currents flowing in the copper coils. Some, like those at Oak Ridge, operated steadily, using direct current (DC), as reflected in names such as "Direct Current Experiment" (DCX). Some were cooled by water flowing through hollow paths through the conductors, and others were cooled by liquid nitrogen to reach lower temperatures and reduced resistance. Most were merely "pulsed" on for a second or less to allow for radiative cooling between "shots." In some cases, choices also reflected the availability of electric power, Oak Ridge having access to the vast power resources of the Tennessee Valley Authority, while even today Princeton operates its TFTR only for seconds per shot, storing up energy between shots in a huge flywheel spun up slowly by electric motors of moderate power.

In the first book on fusion energy, published in 1959, Albert Simon (my first boss at Oak Ridge) assumed, quoting Princeton reports, that a toroidal fusion reactor such as the stellarator would employ copper conductors for all of its coils, and that the reactor would have to be very large just to produce enough fusion power to operate the coils. While all this would soon change with the advent of commercially available superconducting magnets that consume little or no electric power, the actual use of superconducting magnets in tokamak research was considerably delayed in order to postpone the added complexity of cooling the coils to the extremely low temperatures required. Thus, even though several superconducting mirror devices were operational before 1970, all of the large tokamaks operating today—including TFTR, JET, JT-60, and DIII-D—utilize conventional copper coils.

The early interest in DC superconducting mirror experiments was motivated to a considerable extent by the fact that the neutral beams then available were very weak and therefore required a long time to accumulate a dense plasma. Whereas the plasma current alone could produce interesting plasmas for study in the early tokamak, magnetic mirrors required some means of heating the ions to confine them, since at low temperatures the ions escaped very quickly as a result of collisions. Thus it was a perceived necessity that caused mirror researchers always to be at the cutting edge of the development of new technologies for heating and for magnets.

At first thought, the idea of using superconducting magnets in a magnetic fusion reactor is simply mind-boggling. On the one hand, we require a plasma at a temperature of at least 100 million degrees Celsius, even hotter than the center of the Sun. On the other hand, we must surround this fantastically hot gas cloud with several thousand tons of magnets that, to maintain the superconducting state, must be continuously cooled to a temperature of minus 269 degrees Celsius, just 4.2 degrees above absolute zero—about like the cold of outer space.

However, recalling that the low-density plasma has relatively little heat energy despite its fantastic temperature, we would find that it is not the high temperature of the plasma but rather the enormous fusion power it produces that the magnet engineer must reckon with—several billion watts in a full-scale fusion reactor. Removing this heat from the vacuum vessel and the components outside the vessel requires serious engineering, but nothing beyond the experience of other electric generating systems that must continuously conduct away billions of watts of thermal power through flowing water or other coolants. But because the refrigeration required to cool the magnets is very inefficient, even a small leakage of heat into the magnets would demand millions of watts of additional cooling power, too expensive and wasteful to be tolerated. The solution is to place the magnets outside all other components, which then form an insulating barrier a meter or so thick, which largely stops both heat conduction and radiation of all kinds.

Though superconductivity was discovered by H. Kammerlingh Onnes in 1911, the quantum mechanical explanation of the phenomenon was first published by John Bardeen, Leon Cooper, and J. Robert Schrieffer in 1957; this explanation was soon followed by the development of practical superconductors. Metals such as copper conduct electricity in the form of the motion of individual electrons in the metal, resistance being due to collisions of these electrons with the metal ions. By contrast, superconductivity depends on a "pairing" of two electrons, via their coupling to the metallic array of ions ("lattice interactions") in such a way that the electron pairs cannot give net energy to the ions, hence the absence of resistance. Pairing is a quantum-mechanical effect, immune to the collisional resistance experienced in ordinary conductors. However, this pairing effect is easily destroyed by thermal energy, which is the reason that most materials capable of becoming superconducting must be very cold to do so. A very high external magnetic field or a high average current density in the conductor also destroys superconductivity. Thus, the practical use of superconductors in magnets producing the high fields of interest to fusion had to await the development of materials that could remain superconducting at these high fields. By the 1960s, this began to happen with the advent of conductors made of niobium metal alloyed with tin or titanium. More recently, ceramic materials have been discovered that become superconducting at much higher temperatures, but thus far it is not clear whether they will be of practical use for fusion.

Having already successfully operated the Baseball II superconducting mirror facility, Livermore became more deeply involved with superconductivity with the decision, following the exciting results on 2XIIB, to proceed with the construction of the Mirror Fusion Test Facility (MFTF, later MFTF-B). The largest component was to be a superconducting magnet about five meters across, designed to produce the "magnetic well" we encountered in chapter 5. The design, in two pieces, was called the "yin-yang," after the Chinese symbol of opposites.

The main component requiring engineering development was the conductor. Typical of superconductors, it consisted of strands of niobium-titanium alloy embedded in copper that serves as a "stabilizer," the idea being that the copper would take over carrying the current if a transient rise in temperature caused the niobium-titanium strands to lose their superconductivity temporarily. The manufacturing process started with a short, fat sandwich of copper and niobium-titanium wires that was then extruded, or stretched, to form a long conductor whose cross-section was about one centimeter square. The conductor was to be cooled with liquid helium. (Helium liquefies at a temperature of minus 269 degrees Celsius.) The conductor was wrapped in a copper jacket perforated to allow the liquid helium to circulate.

Because all of this was new to industry at that time, Livermore engineers assisted in supervising the conductor fabrication work actually carried out by an outside company, and they themselves undertook the process of winding the thirty-one miles of stiff conductor onto forms to provide the proper "yin-yang" shape. Winding the coil took about a year. In a later modification of the machine that became the much larger MFTF-B "tandem mirror" system, all of the coils that were subsequently added, including one more yin-yang coil, were manufactured in San Diego by General Dynamics, again with Livermore assistance where needed. From this experience, General Dynamics went on to become a manufacturer of superconducting magnet systems, including magnetic resonance imaging (MRI) medical diagnostic devices.

At last, in early 1982, the first yin-yang magnet was ready for testing. At that time, it was the largest superconducting magnet ever constructed. It weighed 350 tons, like a Boeing 747 fully loaded. Most of the weight consisted of high-strength steel reinforcement to withstand the enormous forces to be expected when current was turned on in the coils. For the tests, it was necessary to move the magnet several hundred yards from the winding site to its final location in the seven-story building that would house the MFTF experiment. Since it was far too heavy for cranes and trucks, the engineers finally resorted to the ancient Egyptian method of rolling the magnet forward on a bed of logs, and moving logs forward as the magnet rolled over them. We mounted photographs of this marvelous undertaking in plastic paperweights as souvenirs.

During testing, after the current in the yin-yang was carefully increased step by step, signals in the control room indicated a possible fracture in the yin-yang's steel reinforcement. Theodore Kozman, the engineer in charge, brought me the facts for a decision. The yin-yang magnet had cost about $15 million, so we dared not ruin it. Yet the situation was very much like the problem that an airline pilot faces when he has a red light on his instrument panel. The problem might be real, or the instrument might be faulty. While the Federal Aviation Administration has strict procedures for use in such cases, Kozman and I had to decide for ourselves. Based on all that he had seen thus far, it was Kozman's guess that the problem was "instrumental"—a break in little resistive signal wires taped to the magnet case. It would be easy to see this if we could take a look at the wires, but we could not just walk right up to a huge magnet at the temperature of outer space, and warming the magnet up would take a week. By the next day, we decided to go ahead on the grounds that if there were a crack we needed to find out, so that we could reweld it. The risk was that we might damage the structure too much before we were sure it had ruptured, but we concluded that this was unlikely. As it turned out, the test was fully successful all the way to the design current, and indeed the little wires had simply come undone. Since this came at a time when a superconducting accelerator project elsewhere was having great difficulty, our success helped restore faith in superconducting magnet technology in the U.S. Department of Energy.

Suspenseful moments like this are part of a day's work in Big Science projects. Sometimes progress is not as smooth as you would like, as when the first baseball magnet at Livermore suddenly burst apart, popping a bolt head the size of a football helmet; or the day a huge coil was dropped from a crane during the construction of TFTR. They matter most when, despite all precautions, a life is in danger. On the day of a sizable earthquake in January 1980, one of our technicians was working high atop the yin-yang vacuum vessel about seventy feet above the floor. Fortunately, though very much frightened, he was secure in a safety harness. A physicist accidentally electrocuted in my early days at Oak Ridge was not so lucky.

Besides neutral beams and superconducting magnets, another ITER technology first developed for the special needs of magnetic mirror research is "electron cyclotron heating," the heating of electrons by intense microwaves. This work was pioneered at Oak Ridge by Ray Dandl and others when I still worked there in the early 1960s, so I had the opportunity to participate as a theorist. The motivation for mirror scientists to develop electron cyclotron heating, as well as neutral beams, lay in the fact that magnetic mirrors can only confine energetic particles, so that some means of rapidly bringing particles to high energy is required. In neutral beams, ions acquired the necessary energy by being first accelerated outside the mirror device and then transported, as neutral atoms, into the plasma. In electron cyclotron heating, electrons already inside would be rapidly accelerated by absorbing microwave power, as occurs in a cyclotron accelerator, hence the name. It was Dandl's hope that electrons in a mirror device could in this way be heated to such high energies that their leakage out of the mirrors by collisions would become irrelevant. The ions would then be confined by the electrostatic charge of the electrons, which would be confined in turn by the magnetic mirrors.

There were two other motivations for embarking on the development of electron cyclotron heating. One was the fact that microwaves could be transported for long distances and around corners, and thus electron cyclotron heating would make the task of getting power into the magnetic confinement device easier for designers than did the energetic ion beams then in use at Oak Ridge, or even the later neutral beams, which required line-of-sight access for the beamlines through sizable entry ports. This feature would also turn out to make electron cyclotron heating attractive for tokamaks. The other reason had to do with the availability of high-power microwave sources developed for military radar installations. Later, by the way, this would become a point of some contention between us and our Russian colleagues when the Soviet fusion authorities were not allowed to share with us information about their version of a high-frequency microwave tube called a gyrotron, though they had access to the American version.

Microwave heating depends on the phenomenon of resonance. To understand this, recall that we learned earlier that electrons moving in a magnetic field spin in circles at a rotation frequency that depends on the strength of the field. As they do so, they emit electromagnetic radiation as if each spinning electron were a current oscillating in a tiny antenna the size of the electron gyroradius. This is the principle of the "magnetron" in your kitchen microwave oven, and the high-frequency version called a gyrotron. If the gyrotron is connected to a tokamak by means of a tube of the proper cross section, known as a "wave guide," and if the magnetic field in the gyrotron is the same as that in the tokamak, any cold electrons in the tokamak will begin to spin by absorbing energy from the microwaves.

Because the two magnetic fields in the tokamak and the gyrotron must be the same, electron cyclotron heating is also called "resonant heating." Two common experiences of resonance are tuning your radio (to "resonate" with the station) and pushing a child's swing. If you push just when the swing is ready to go forward again, you do the most good. Similarly, the microwaves work best if their frequency (just the electrons' frequency of spin in the magnetic field of the gyrotron) matches the frequency of spin in the tokamak, both frequencies being determined by the magnitude of the respective magnetic fields. A mismatch in frequencies by exactly a factor of two ("second-harmonic resonance") also works.

Resonant heating works for ions as well as electrons, using frequencies in the shortwave radio range tuned to match the frequency of spin of the heavier ions. The JET has twenty megawatts of such ion cyclotron heating power, as well as neutral beams. The largest electron cyclotron heating system for tokamaks is that for the T-10 tokamak in Moscow (the seventh generation beyond T-3, which turned the world's attention to tokamaks back in 1968). Other resonant frequencies are also utilized, such as the so-called lower hybrid frequency lying between the ion and electron spin frequencies.

The relative ease of transferring heating technologies between mirrors and tokamaks is due to the fact that they are auxiliary to the magnetic containment device. Like the diagnostic instrumentation described in the previous chapter, a heating technology could often be added to a tokamak machine even as an afterthought, long after the machine was built and operating, provided only that there was space in the building for the equipment and an access port available to allow entry of the new system into the vacuum chamber. Indeed, in almost any fusion laboratory of the past, time-lapse photography over a few years would disclose an ever-increasing complexity of auxiliary systems, even to the point of hiding the confinement device itself, as new heating and diagnostic equipment was invented or adapted from successful applications elsewhere. A mature device such as ITER will have space for a number of auxiliary systems provided from the start. I still recall my sense of wonder, in our own MFTF mirror facility, as I wandered from floor to floor and room to room among the huge electrical power supplies that converted the 150 megawatts from our electrical substation (enough for a small city) to the voltages and currents required to operate the MFTF eighty-kilovolt neutral beams and gyrotrons.

The flexibility for change and modification among auxiliary components does not apply when we turn to the magnetic confinement device itself, especially the tokamak, which is a highly integrated structure, beginning with the toroidal field coils. Thus, while the big tokamaks initiated in the 1970s did not use superconducting magnets, direct experience with the construction of superconducting toroidal field coils was needed, and an international program was soon established for this purpose. This was the Large Coil Test Facility at Oak Ridge, which in 1985 tested six large toroidal field coils, three provided by the United States, two from Europe, and one provided by Japan. The dimensions of the coils, which had an inner bore of 2.5 by

3.5 meters, were somewhat smaller than those of the copper coils for JET, and therefore considerably smaller than the linear dimensions of the ITER toroidal coils. Nonetheless, the fact that these coils were successfully tested more or less without incident was encouraging. The purpose of testing the six coils at once was to utilize the combined magnetic field they generated to simulate the very strong field that exists at the center of a tokamak where the toroidal coils come together. The magnetic field strength, in gauss units, must exceed 120,000 gauss at this point in order for there to be a 50,000-gauss field at the center of the plasma, as in TFTR and in the ITER design. (For comparison, the natural magnetic field at the Earth's surface is about 0.33 gauss.) In addition to these tests, a superconducting tokamak the size of TFTR is now in operation in France. This is the Tore Supra tokamak at Cadarache.

While most of the early technology development was initiated by various research groups to meet their own needs, by the late 1970s, both in the United States and elsewhere, development for future needs began to be organized nationally, or even—as in the case of the Large Coil Test Facility project—internationally. Correspondingly, technology development broadened to include new research areas of interest to engineers looking beyond plasma experiments and toward the technologies that fusion systems would ultimately need if they were to be a viable energy source. Thus work began on diverse topics such as the development of new materials that could withstand heat and radiation, the safe handling and processing of radioactive tritium, and conceptual designs of commercial power plants. A focal point of this engineering research became the Fusion Materials Irradiation Test (FMIT) facility at Hanford, Washington. Predictably, as budget pressures mounted in the United States in the 1980s, the FMIT was the first fusion project to be canceled, in order to preserve funds for TFTR and other plasma physics facilities. However, by 1988, a new focal point for technology development arose with the formation of the ITER project, this time on a thoroughly international basis.

10 : The International Thermonuclear Experimental Reactor

The real internationalization of U.S. magnetic fusion research can be traced to the arrival in Washington in 1981 of Alvin Trivelpiece as director of research in the U.S. Department of Energy. This was soon after the passage of the Magnetic Fusion Engineering Act of 1980, signed into law by President Carter shortly before leaving office. This act, never implemented, had called for large increases in fusion appropriations, and the construction by 1990 of a fusion Engineering Test Facility something like what later became the International Thermonuclear Experimental Reactor (ITER), but run and funded by the United States alone.

As a new political appointee in the Reagan administration, Trivelpiece saw things very differently. While the concept of "energy independence" for the United States had first been floated under Republican presidents, Nixon and Ford, the new Reagan White House seemed determined to eliminate energy from the federal agenda, even to the extent of abolishing the new Department of Energy, which President Carter had counted among his finest accomplishments. I was keenly aware of this apparent sea change in U.S. politics. In the fall of 1980 I had seen a "white paper" from the Reagan transition team calling for the abandonment of energy independence, and reliance instead on the international marketplace for oil. Dismayed, I had mentioned this white paper to an aide of Senator Paul Tsongas at a breakfast meeting for fusion leaders in Washington, but at the time I was unable to remember the author's name. Later the aide called to say that the author of the white paper, David Stockman, had just been appointed to head Reagan's Office of Management and Budget.

As Trivelpiece told me later, immediately upon arriving in Washington he had called upon the president's science adviser and the head of the National Science Foundation to size up the situation for research. From these meetings and his own analysis, he had become convinced that the only hope for maintaining a viable magnetic fusion program in the United States—and indeed, the world—was in pooling our talent and resources in a fully coordinated international endeavor. I soon found myself among his willing accomplices in this enterprise. I recall in particular a meeting on international science projects that Trivelpiece and I attended in Copenhagen. There I learned that throughout the world Big Science was grappling with the same questions of how to move forward with costly projects, and that, like Trivelpiece, most science administrators suspected that in the future individual nations would no longer be able to handle most such projects.

Though his duties were wide ranging, Trivelpiece had arrived in Washington with especially strong credentials to lead fusion in new directions. He was a former professor of physics at the University of Maryland and coauthor of a well-known textbook on plasma physics, and he had been a colleague of Robert Hirsch in the Office of Fusion Energy during the heyday of fusion program growth in the 1970s. However, not only did he need to win over his fusion colleagues to implement international cooperation on fusion research, he also needed the right political framework. Despite a temporary estrangement from our Russian fusion colleagues after a Soviet military intervention in Poland, he soon found that framework in President Mitterrand's Versailles summit of the Western industrialized nations and Japan, one of the annual economic summits of the Group of Seven. Mitterrand's agenda was heavily oriented toward technology development, with fusion prominent on the list.

While I and many others attended numerous meetings in the United States, Europe, and Japan over the next five years, the ambitious internationalization project made only small headway against the countercurrents of national proposals that still won the greater loyalty of our colleagues back home. Meanwhile, the rise of General Secretary Mikhail Gorbachev to power in the Soviet Union presented a new and unexpected opportunity. Traveling with Gorbachev on his peacemaking journey to England in 1985 was Evgenii Velikhov, a fusion scientist who had risen to political prominence as an elected member of the Supreme Soviet and vice-chairman of the Soviet Academy of Sciences.

I should digress at this point to comment on relations between American and Soviet fusion physicists during the cold war. While some American physicists boycotted their Russian colleagues as a protest against Soviet human rights violations, many of us in fusion research chose to maintain our relationships, some of which went back to 1958, as open channels of communication. In the end, both approaches played a role in easing the strictures on dissident Soviet physicists, both prominent and obscure, and in aiding a smoother transition for scientists than might otherwise have been possible during the recent tumultuous changes in the Russian political system. Our Soviet counterparts also faced such choices. On more than one occasion, Soviet colleagues explained to me their stance on doing what good they could inside the Soviet system. I do not mean to exaggerate the role of fusion in all this, but fusion scientists have played their part, even in trivial bureaucratic ways. For example, one evening at the U.S. Embassy in Moscow in the early 1980s, the U.S. science attaché explained to me that it was only the arrival of our fusion delegation, after yet another U.S.-Soviet political hiatus, that had allowed him to resume contact with his Soviet counterparts so that he could get on with his work.

Returning to the events of 1985, the first summit meeting between Reagan and Gorbachev finally provided the vehicle that Trivelpiece had been seeking to launch a truly international effort in fusion. The initiative came from the Soviets through Academician Velikhov, acting as adviser to Secretary Gorbachev, and was assisted by Trivelpiece, in the Department of Energy, working through the secretary of energy and the Department of State. When President Reagan addressed Congress about the summit, the only technical item among his five points was a U.S.-Soviet agreement on a major step in fusion research, through international cooperation. This was the birth of ITER.

With ITER, the many years of Soviet-American cooperation in fusion paid off magnificently. I am sure that this could not have happened without the close personal relationships that had grown through all those years. Like many of us, Trivelpiece had known Velikhov throughout the latter's meteoric rise. I myself had met Velikhov in 1964, and I recall vividly Lev Artsimovich's visit to Livermore to let us know that Velikhov would succeed him as the leader of fusion research at the Kurchatov Institute. Even as his career expanded, Velikhov had kept a close eye on fusion, attending personally the many committee meetings I had also attended in the Soviet Union. Some of these were meetings with Trivelpiece to arrange each year the formal exchange visits of scientists between the United States and the Soviet Union. During one such visit, on a flight to Sukhumi (near the Black Sea) that suddenly took a nose dive, Trivelpiece, once a pilot, turned to me, white-faced, to inform me that we had just experienced a flameout.

Just as the ITER project was taking shape, the Lawrence Livermore National Laboratory was being forced by declining national fusion budgets to abandon mirror research. Trivelpiece approached me and the Livermore management about taking a leading role in organizing U.S. participation in ITER. Thus by 1987 I found myself in the position of principal U.S. representative in a working group to establish technical goals and objectives for the ITER tokamak project. To my great relief, Paul Rutherford from Princeton had joined me in this, and I was fortunate in organizing an able committee of advisers at home from many U.S. laboratories and universities. My principal Soviet counterpart was Academician Boris Kadomtsev, whose theory of turbulent transport we encountered earlier and who now headed fusion research at the Kurchatov Institute. At the insistence of the United States, our Japanese and European partners from the "Versailles process" had been invited to participate as full partners in ITER, the European nations acting as a single group through the European Community (now called the European Union) in Brussels. The project was to be organized under the auspices of the International Atomic Energy Agency (IAEA) of the United Nations. All of the meetings of the working group were held at IAEA headquarters in Vienna.

From the outset, there was little doubt that the working group would succeed. We all knew we were there to make it happen, especially the Soviets and the Americans. Among all of us there was, I think, an exciting sense that, though there had been some false starts in the past, ITER was something new. This was not the first attempt at organizing a fusion engineering test facility on an international basis. Many of the individuals present had been involved to some extent in the INTOR design project, now aborted, which had also first been advocated by Velikhov and the Soviets through the IAEA; and an engineering test facility had been the unrealized long-term goal of the Versailles partners. There was a strong desire not to repeat this history, not to have "just another paper study." On the other hand, by now both the Europeans and the Japanese were designing their own large national fusion projects, just as the United States had done earlier, and Princeton was pushing for a more modest project, called the Compact Ignition Tokamak (CIT), though we contended that this did not compete with ITER. Only the Soviets were not seriously pursuing a home project.

My job was to help bring about a consensus on ITER as expeditiously as possible. English being the working language, it often falls to the Americans in such international science gatherings to draft the words that all can agree to, and this was to be no exception. The name of the project, by the way, while it is an acronym, also, fittingly, means "the road" or "the way" in

Latin. But our Japanese colleagues said that it is also close to their word for "big headache."

Most of our headaches were technical, though it was also our job to suggest guidelines for organizing the project, assigning intellectual property rights, and so on. Any truly political issues were to be handled by the appropriate governmental representatives. One interesting political side-issue concerned the fact that this was the first instance in which the Soviet government was dealing directly with the European Community as a political entity. How, then, to handle countries such as Switzerland, which were not member nations but were important affiliated players in fusion? And how to do this without embarrassing the Soviets, who had no desire to undertake similar negotiations for all of their own friends in the Warsaw Pact? Fortunately such matters were someone else's headache.

The main technical point of contention was whether ITER's primary goal was to serve as an engineering test facility that would advance the nuclear engineering of fusion as a practical power source, or to demonstrate ignition with DT fuel to prove that the tokamak was capable of sustaining a fusion reaction by itself with little or no external source of power. Either goal would have been a major advance beyond TFTR and JET, which at that time had not yet ventured into operation with actual DT fuel; and beyond the JT-60, which, though as large as JET, was not designed for DT operation. However, there was a strong desire, especially in the United States at that time, that ITER should fulfill both goals in a single device, since that was the only objective of sufficient scope to justify a large international enterprise. The Europeans were the most insistent that the ignition goal must drive the design, though in fact everyone wanted ITER to achieve ignition.

The resolution to the problem, soon arrived at, was a two-phase approach. The first phase was to be the achievement of sustained ignition for a significant period of time, hundreds of times longer than shots in TFTR and JET. The second phase was to focus on engineering testing of "blanket" modules, described below, that would be prototypical of the nuclear engineering equipment needed to turn the kinetic energy of fusion neutrons into usable heat energy for the generation of electricity. To prepare for the engineering phase, the machine would be designed to permit continuous operation for specified, long periods of time at power levels adequate to yield meaningful nuclear engineering information; and radiation shielding was to be included in the design to allow operation at such power levels over a number of years. Consensus on these and other details was reached through

a series of meetings, and through much work and discussion at home between meetings. The working group finished its task by 1987, in time for formal negotiations to set the ITER project in motion by April 1988.

Like the project itself, the ITER design activity was divided into phases,
and each phase was to conclude with a decision on whether to continue the
project. The first phase, which began in 1988, was called the Conceptual
Design Activity, and was scheduled to be completed by December 1990.
The main work was to be carried out at the German fusion center at
Garching, near Munich, by a team of about fifty scientists and engineers divided equally among the four partners. Since the site was European, the
other partners had priority when leadership roles were assigned. Thus, Ken
Tomabechi, of Japan, directed the design activity in Munich; John Clarke,
director of the U.S. magnetic fusion program, headed the overall governing
body, called the ITER Council; and Boris Kadomtsev, of the Soviet Union,
chaired a twelve-person scientific and technical advisory committee that reported to the council. I sat on Kadomtsev's committee along with two other
Americans—Paul Rutherford, of Princeton, who had served with me on the
ITER working group, and Robert Conn, then a professor at UCLA and the director of its Institute for Plasma and Fusion Research. Sometimes John
Clarke asked me to sit in on ITER Council meetings, also. By this time, I had
left Livermore to become a professor in and chair of the Department of
Nuclear Engineering at Berkeley.

As the conceptual design progressed, some of the divisive issues dealt
with earlier by the working group began to resurface in our technical advisory committee meetings. Again I found myself trying to settle differences,
often between ourselves and our European colleagues, as Tomabechi and
the conceptual design team struggled to meet the ITER goals within budget.

To help the reader appreciate what was at stake here, let us think back to
some of the points we learned about tokamak design in earlier chapters. In
chapter 2, we learned that the tokamak consists mainly of a big toroidal tube,
or vacuum vessel, to hold the fuel plasma; a solenoidal magnet, to produce a
strong toroidal field around the tube; and some means of driving an electric
current in the plasma. All these features, shown in primitive form in figure 1,
reappear in the actual ITER conceptual design drawing (fig. 4). In addition,
in ITER, the vacuum vessel must be surrounded by a thick shell, or blanket,
a meter or so in depth, which contains the coolant fluid that removes the
enormous heat to be generated, especially in the engineering phase, and also

(as we learned in chapter 9) provides the shielding that is essential to protect the superconducting magnets from radiation.

The point of contention lies in the fact that all three of the major components—the vacuum vessel and blanket, the toroidal field coils, and the current-drive system—compete for space in the center of the machine. In other words, as is clear from the ITER drawing, the ITER tokamak is a torus, or "doughnut," that is crammed full of equipment in the middle. And the more equipment designers must place in the device's middle, the bigger the machine becomes overall, all dimensions swelling in proportion if we adopt the conservative approach of simply scaling up ITER from the earlier successful tokamaks such as JET or TFTR.

When assigning priorities to the three major components, there is no getting around the need for the toroidal field coils, which most researchers would agree must be superconducting in a machine of the size and purposes of ITER. Since superconducting coils demand radiation and heat shielding, and all parties want ITER to produce large amounts of fusion power, there has also been little debate in principle about the need for a protective blanket, even in the center of the machine. And since a stronger toroidal field permits a higher plasma pressure and fusion power (see chapter 2), all agree that the toroidal field coils must be big enough and sufficiently reinforced with steel to produce the highest fields possible. Thus the main component at issue has been the current drive system.

Scientists at Livermore had become interested in current drive when we first became involved with ITER and tokamaks, back in 1986. Up to that time, all tokamaks had relied solely on the inductive method of current drive, which requires an air-core transformer coil (labeled *central solenoid* in fig. 4) to drive the current. Since the inductive current drive coil occupies a significant fraction of the precious space in the middle of the machine, the Livermore team began to investigate the benefits of eliminating this coil.

By this time, tokamak scientists had actually demonstrated other, noninductive means of driving the current using auxiliary heating technologies that can be applied from the outside (see chapter 9), without involving the middle of the tokamak. While both neutral beams and microwaves can drive the plasma current, the best-understood method is a neutral beam oriented tangentially to the plasma torus so that ions deposited by the beam acquired a large component of velocity parallel to the twisting field lines. Since electrons trying to follow the newly injected fast ions would cancel the current, the actual generation of a new current depends on collisions between electrons and slower impurity ions already present, as first pointed out by Ohkawa in a paper published in 1970. The Livermore design studies confirmed that eliminating most of the inductive current drive capability would make ITER much more compact and also permit steady-state operation, which most engineers considered more desirable for future reactors than the on-and-off pulsation of a transformer-driven machine. A big drawback is the fact that a very high-power neutral beam would be needed to drive all of the current, and such a beam would consume too much electricity. To offset this power demand, the Livermore team assumed that the bootstrap current, not yet confirmed experimentally at that time, would drive a significant fraction of the current.

While noninductive current drive and the bootstrap current have now been demonstrated in the laboratory, in 1986 I quickly concluded that a full reliance on noninductive current drive was too advanced to be accepted as the design basis for ITER. Thus, by the time the ITER working group was convened in 1987, I knew that the United States must accept a substantial inductive current drive capability in the design, though we did insist that development of noninductive current drive be pursued as part of the ITER project and that the flexibility be maintained to add noninductive current drive to ITER later.

Predictably, the size and projected cost of ITER increased with the addition of substantial inductive current drive. In the final conceptual design report, issued in December 1990, the major radius of the device was 6.0 meters, about double our starting point at Livermore, and the current was now 21 million amperes, nearly double the current in the Livermore case. The inductive current drive was intended to be sufficient to build up the current and sustain it for three hundred seconds. According to my own rough estimates, these additional capabilities had been gained at a cost of roughly $1 billion per additional meter of major radius.

Current drive was still very much an issue when the next phase of the ITER project, the Engineering Design Activity, got under way in June 1992. The new project director for this phase was Paul Rebut, who was to serve as ITER director from 1992 to 1994.

Rebut had been the force behind the creation of the JET facility in Europe and was serving as its director at the time he was appointed to take charge of the ITER engineering design phase. A highly respected scientist and visionary, Rebut had strongly influenced the design of JET, and he now brought his ideas to ITER. By this time I no longer had a direct role in ITER, but I watched with interest as Rebut took hold of the project. Because we had served together on Kadomtsev's advisory committee for the conceptual design phase, I knew that Rebut was very dubious about noninductive current drive and was determined to increase the inductive capability of ITER. It was no surprise, therefore, that under his leadership the inductive current drive capability was increased to a thousand seconds. This and other engineering changes increased the major radius another two meters.

Besides focusing on current drive during his tenure as director, Rebut and the ITER team also gave considerable attention to another tokamak technology component of particular interest to me. This is the so-called divertor, which was left in a somewhat unsatisfactory state at the conclusion of the Conceptual Design Activity. The purpose of the divertor, not yet discussed in this book, is to carry away the enormous power deposited in the plasma by the alpha particles produced by fusion reactions. While some of this power is radiated by electrons as x-rays and microwaves, most of it diffuses across the magnetic flux surfaces in the form of plasma still hot enough to transport power. Experience has shown that, uncontrolled, all of this plasma power is likely to concentrate in some unpredictable location, with consequent release of gaseous and metallic impurities wherever the plasma happens to land on the surface of the vacuum vessel. The function of the divertor is to channel this plasma power to a specific region where unwanted impurities can be pumped away and coolants can be provided to remove the heat. This is accomplished by modifications to the magnetic field such that the outermost field lines do not form closed flux surfaces but instead are "diverted" to intercept structures designed to handle the power. The term "divertor" refers to these structures, and to the modifications required to divert the field lines.

In ITER the field lines are diverted by means of circular "poloidal" magnet coils above and below the tokamak, shown in figure 4. The divertor structures themselves are labeled *divertor* in figure 4. The poloidal coils serve several interrelated functions. First, they create a field in the vertical direction that prevents the toroidal plasma from expanding in major radius (a problem mentioned in chapter 2). Second, by properly adjusting the currents, these coils stretch the plasma cross section into an elliptical shape that allows more plasma current to flow in a machine of given size. Without this feature, ITER would be even larger. And third, to create the divertor, these coils define a boundary inside of which the plasma current creates closed flux surfaces (see chapter 5), while outside this boundary the field lines, dominated by the external poloidal coils, turn upward (or downward) toward the wall of the vacuum vessel. The recessed chamber where these diverted field lines intercept the metal wall comprises the divertor structure. Plasma that diffuses across the boundary between the closed flux surfaces and the diverted field lines flows into the divertor structure.

Even though the plasma temperature in the divertor region is not high, a mere million degrees Celsius or less, the plasma flow rate into the divertor is sufficient to conduct away all of the plasma power that is not radiated. In ITER, this means that a sizable fraction of three hundred megawatts, in the form of flowing plasma heat, is headed toward metal plates in the divertor structure. No matter how one tilted these plates to present the greatest surface area, they would be doomed to receive heat loads exceeding ten megawatts per square meter—far more than coolants could handle. Faced with this, the conceptual design team had been unable to find an operating regime in which steady-state noninductive current drive and safe divertor operation were mutually consistent at the full power levels required for the engineering phase.

The engineering design team began to examine a different approach based on new ideas emerging from experiments on the DIII-D and elsewhere, the DIII-D and the JT-60 being the only large tokamaks equipped with divertors. The basic idea is to get rid of the heat by puffing gas into the path of the plasma streaming toward the divertor plates—a process analogous to blowing out a fire, except that in this case the heat is radiated by the gas, not conducted. At this writing, though the approach is promising, everything is not understood about this "gaseous divertor." In fact, one of my graduate students has worked on divertor models in collaboration with professionals involved in the DIII-D program. An important question, not fully answered, is whether or not the gaseous divertor will be mutually compatible with plasma confinement in the so-called H-mode, described earlier, which depends sensitively on plasma properties in the colder "edge" near the boundary of the closed magnetic flux surfaces. Physics at the plasma edge has become a field unto itself.

The ITER Engineering Design Activity, now under the overall direction of Robert Aymar, is expected to continue until 1998. Activities are carried out in three design centers—in San Diego; at Garching; and at Naka, the site of the JT-60. Because of economic problems, the Russian Federation, which replaced the Soviet Union as an ITER partner, does not maintain such a site, but the Russians have been faithful to their commitments to provide scientists and engineers to work at the other sites. As in the previous phase, all sites are international: a Russian heads the American center, an American heads the Garching center, and a European heads the Naka center. Because the project is spread around the world, "the sun never sets on ITER." All the centers have extensive interconnections through electronic communications.

About 150 scientists and engineers are engaged in these design activities, roughly four times the number involved in the conceptual phase, and the engineering design phase is intended to last three years longer than the conceptual design phase. In addition to the engineering design work, this phase includes a considerable amount of technology development work to be carried out by each of the partners under "contract" with the ITER director, in the sense that the tasks must be mutually agreed upon in order that the partner receive credit for the work as part of its agreed-upon financial contribution to the project.

The status of the ITER engineering design, at this writing, is given in the table on page 124, which lists a few key parameters. Also listed for comparison is the same information for TFTR.

The overall height of the ITER machine, roughly five times the plasma major radius, would be around forty meters, or 130 feet, about twice the height of the MFTF vacuum vessel we built at Livermore. Actually, the containment vessel for a nuclear fission reactor producing only twice the thermal power of ITER is typically higher—one California example that comes to mind is some two hundred feet high. At first, fusion physicists used to smaller equipment had a hard time grappling with such numbers. I am reminded of an occasion in the late 1960s when, to help educate them, a group of fusion experts attending a conference in England, still naive about industrial-scale projects, was reportedly taken on a tour of a nearby coal-fired electric power plant. As I heard the story, without explanation the group was led, wide-eyed, into an enormous empty building lined with bricks and pipes. "Friends," the guide announced, "you are now standing in the boiler."

Clearly, ITER would be a major step forward from where we are today. Fifteen hundred megawatts of sustained fusion power in ITER is already half the thermal power of the largest fission reactors now operating. More to the point, the leap from a one-second pulse of ten megawatts of fusion power in TFTR to fifteen hundred megawatts of sustained fusion power in ITER sounds like a technological fairytale. Yet, as I have tried to share with the

ITER Design Parameters (in 1995) and TFTR Parameters

	TFTR	ITER
Plasma radius (meters)	2.5	8.1
Magnetic field (tesla; 1 tesla = 10,000 gauss)	5.2	5.7
Plasma current (millions of amperes)	2.5	21
Neutral beams or other auxiliary heating power (millions of watts)	40	100
Fusion power (millions of watts)	10.7 (peak)	1,500 (more or less continuous)

Sources: TFTR data from Weston M. Stacey Jr., *Fusion: An Introduction to the Physics and Technology of Magnetic Confinement Fusion* (New York: Wiley-Interscience, 1981); ITER data from ITER Interim Design Report, IAEA/ITER EDA/DS/07, July 12, 1995, obtainable through the International Atomic Energy Agency, Vienna.

reader in the foregoing chapters, far from being the creation of wild-eyed optimists, ITER is better characterized as the conservative effort of a worldwide community of tokamak scientists and engineers determined that their record of accomplishments, which has seen no real failures over more than twenty years of working at the frontier of knowledge, remain failure-free.

The goals of the ITER project remain essentially as they were defined by our working group in 1987—first, to demonstrate ignition or sustained burning; and second, to provide a testbed that all partners can use for the development of nuclear engineering components, such as the blanket.

Edwin Kintner, the nuclear engineer–physicist who succeeded Hirsch as director of the U.S. fusion program from 1976 to 1981, liked to point out that a great advantage of fusion was its total separation of the power production in the blanket from the energy source in the plasma, whereas these functions are thoroughly commingled in a fission reactor. It is this separation of functions that makes a fusion blanket test program credible. In other words, although ITER might not turn out to be an exact prototype of a commercial electric power plant and was never meant to be, it would nonetheless provide the same kind of fusion energy source that any other future plasma device would provide, and this fusion energy source could be used here and now to operate and test blanket sections of any design. In this way, nuclear engineering could proceed in parallel with improvements in the plasma source. And, of course, the operation of a complete tokamak energy source at essentially full power could only help in advancing knowledge about this particular kind of plasma device. This is the strategy, instituted by Kintner, that first the U.S. magnetic fusion program, and now the world's, has attempted to follow for the last twenty years.

The point of this strategy is that there may be as much to learn about successfully producing thermal power in the blanket as there has been to learn about producing fusion energy from the plasma. And just as plasma experiments are the best way to learn about producing fusion energy, actually designing and testing blankets will stimulate the ideas that make them work best.

To help the reader understand the importance of the blanket, I had better say something about how engineers contemplate using fusion energy to produce electricity. First, while fusion energy, because of its special nature, may eventually find other uses, its earliest application, like that of fission, is expected to be as a replacement for the fossil fuels now used to provide the heat that makes steam (or other hot gases) to drive electric generators. Thus, the most important function of the blanket is to convert the DT fusion energy, 80 percent of which is released in the form of energetic neutrons, to heat at temperatures high enough to produce electricity efficiently—typically, eight hundred degrees Celsius or more. In ITER, it is not essential that the entire blanket operate at such temperatures, but it is essential that the design allow these temperatures in the removable sections of the ITER blanket where researchers from the partner nations will test power-producing blanket designs. Approximately 10 percent of the blanket surface area, roughly one hundred square meters, will be available for these tests—more than enough for all partners, since the blanket of a power reactor would almost certainly be composed of separate modules much smaller than this. As in any other thermal power producer, pipes carrying a coolant fluid such as water, liquid metal, or helium gas would pass through the blanket module, where the fluid would be heated; it would then be transported through the pipes to locations outside the ITER vacuum vessel, where the heat could be used in conventional ways to produce steam.

The second key function of the blanket is the breeding of tritium, which is crucial if DT fuel is to be used. (As we learned in chapter 1, tritium is a radioactive isotope of hydrogen that must be created inside the fusion reactor by bombarding lithium with the neutrons produced by the plasma.) The lithium can be distributed in the blanket in the form of ceramic pellets of lithium oxide, in the form of liquid lithium metal, or in other chemical forms. Since large quantities of tritium are not available, in a society depending on DT fusion for its electricity it would be necessary that the fusion reactor blanket itself be able to breed as much tritium as was being consumed by the plasma, plus a little extra to make up for the radioactive decay of a portion of the tritium into helium-3 before the tritium could be used. The design issues to be investigated include the loss of neutrons by absorption in the mechanical structures required to support the heavy blanket materials, and the diffusion of tritium through the hot blanket material, a leakage that might allow the tritium to escape. Even ITER would have to breed its own tritium in order to have a sufficient supply for the engineering test phase, now called the Extended Performance Phase. However, the initial requirements can be met by tritium produced by existing fission reactors in Canada.

The third function of the blanket, as seen above, is to be a radiation shield to protect the superconducting magnets. Some shielding is obtained naturally from the structures that carry out the other two functions, and the remainder is provided by an additional layer of shielding material outside the blanket proper.

Finally, we note that the hot blanket is the primary safety concern in a fusion reactor, and that neutron irradiation of the vacuum vessel and the blanket's structural and shielding materials is the only source of nuclear waste. Moreover, the ratio of the cost of a blanket module to the power it can handle is an important factor in the overall cost of electric power produced by fusion energy. Thus, the engineering research on the blanket that ITER would surely foster lies at the heart of the environmental, safety-related, and economic characteristics that will determine the public acceptance of fusion energy and its viability in the marketplace.

This concludes our discussion of the science and technology of magnetic fusion. At the time I am writing this, in the summer of 1995, neither the United States nor any other partner nation has as yet made a commitment actually to construct ITER, and sweeping changes in Congress have once again called into question the appropriate level for fusion funding by the United States. We will return to the political context of fusion research in part IV of this book. But first let us catch up with a whole different world of fusion science, in which huge lasers replace magnets, and tiny specks of DT ice could replace a barrel of oil in the twinkling of an eye.

III

ANOTHER WAY: INERTIAL FUSION

11 : The Little Big Bang

The explosion of applications of the laser following the invention of the device in the 1950s is truly astounding. The laser was first proposed in the United States by Arthur Schawlow and Charles Townes in 1958, a working laser was constructed by Theodore Maiman in 1960, and by 1965 the *World Book* encyclopedia was explaining to children, with illustrations and photographs of actual equipment, the many uses of the intensely bright light beams produced by lasers—ranging from powerful beams that could blast holes through steel, to the precisely controlled beams used by surgeons to "weld" in place a detached retina in the eye.

Scientists interested in fusion soon climbed on the bandwagon. Among them was the Soviet physicist Nikolai Basov, who shared with Townes and Alexander Prokhorov the 1964 Nobel prize in physics for the invention of the laser. By 1964, Basov and Oleg Krohkin, and also John Dawson at Princeton, had published papers on the use of high-power lasers to heat frozen pellets of DT fuel to the 100-million-degree-Celsius temperatures required for fusion. At such temperatures the tiny pellet would instantly become a hot plasma that would expand in size at an explosive rate. The larger the pellet, the longer the time needed for it to fly apart. If the pellet was large enough, there would be enough time for most of the pellet material to undergo fusion reactions, producing many times as much energy as was needed to power the laser that heated the pellet. However, it seemed that such a pellet would be so large as to be impractical.

In the meantime, American nuclear weapons scientists, including John Nuckolls, at Livermore, had also been thinking about lasers. In 1958, just six years after the first successful test of a hydrogen bomb, Nuckolls had made calculations for a fusion electric power plant that used the energy supplied by a succession of hydrogen bomb blasts inside a large underground cavity. Nuckolls recognized that to be attractive for commercial power generation, the smallest possible fusion "microexplosions" should be detonated by some means other than the fission bombs used to ignite a hydrogen bomb. In 1959 he used computer programs developed for nuclear weapons design, employing concepts not yet known to physicists such as Dawson, to show the feasibility of igniting a tiny quantity of DT fuel, only a thousandth of a gram (one milligram). Nuckolls's calculation challenged the consensus at

Livermore that the smallest mass of DT that could be ignited was one hundred milligrams, a quantity whose ignition would create a sizable explosion. Thus encouraged, he began to search for a nonnuclear replacement for the fission bomb to power his scheme, and he realized that he had found it when the first laser appeared in 1960.

Nuckolls began applying a variety of weapons design concepts in his computer calculations, now aimed specifically at using lasers to ignite a tiny quantity of DT fuel. Other Livermore scientists, including Stirling Colgate, Ray Kidder, and Ron Zabowski, also addressed this problem. As we shall see, the basic idea employed by all of these scientists familiar with weapons design was that the fuel must first be compressed to a very high density. Other ideas that Nuckolls employed in his calculations were "pulse shaping" and "propagating burn," concepts that are still the essence of laser fusion. In 1960 and 1961, Nuckolls was already able to show that a spherical droplet of "frozen" DT fuel could be compressed to very high densities and ignited by propagating burn from a "hot spot" at its center.

By 1962, just two years after Maiman's first laser demonstration, laboratory director John Foster had asked Kidder to lead a program to build a ruby laser at Livermore. The first step was to be a twelve-beam laser producing a light pulse containing one joule of energy. (A joule is the energy produced by one watt of power in one second.) But it was already recognized that a million-joule laser, probably decades away, would be needed for a commercial power plant based on laser fusion.

It was entirely natural that laser fusion would take root at Livermore. Founded in 1952 by Edward Teller as a center for hydrogen bomb development, the Lawrence Livermore National Laboratory has from the outset also pursued fusion, including magnetic fusion, as an energy source. Teller was a pioneer of nuclear physics and a contemporary of Hans Bethe, George Gamow, and others who developed the compelling evidence that nuclear fusion is the source of energy in the stars. As World War II approached, Teller (who had been born in Hungary in 1908 and educated in Germany, and had emigrated to America in 1935) was drawn into the American development of nuclear weapons—first in 1939, when he helped to solicit Einstein's famous letter that persuaded President Roosevelt to initiate work on the atomic bomb, and by 1941 as an active participant with Enrico Fermi in the bomb project at Columbia University. It was at Columbia that he first became intrigued with the possibility, suggested by Fermi, that the heat of an atomic bomb could be used to ignite deuterium to create an even more powerful explosion. Ten years later, at Los Alamos, Teller conceived the ideas, first applied to hydrogen bombs, that are today the basis for laser fusion as a potential energy source for the twenty-first century. Teller played a major part in supporting the initiation of the laser program at Livermore.

By the time I came to work at Livermore in 1967, laser research was well under way, and similar experiments to study the interaction of laser beams with plasmas were in progress or soon to begin in France, England, Japan, Germany, and the Soviet Union, as well as in the United States at the University of Rochester, United Technologies, Boeing, the Naval Research Laboratory, and the nuclear weapons laboratories at Livermore and Los Alamos.

As a newcomer at Livermore, I began to look around to see what was going on outside my own work in magnetic fusion, and soon became a fascinated observer of the laser research, then a modest effort carried out in secret because of its potential linkage with bomb design. I had arrived at Livermore at the onset of fast-breaking events that would soon catapult laser fusion into a major program at the laboratory. As a member of the wider plasma physics community, I was also aware of the work on laser-plasma interactions being carried out by Dawson and others at other locations. Dawson and I would see each other often at scientific meetings, where we would share whatever was exciting at the time. When on one such occasion he began to tell me things that bordered on what was regarded as secret at Livermore, I invited him to visit with my Livermore colleagues so that they could hear his ideas, with proper security arrangements.

Dawson's visit in late 1968 played some part in stimulating the close collaboration between Nuckolls and Lowell Wood that would change the course of laser fusion research at Livermore. This was only months before reports that Keith Brueckner, a first-rate physics professor at the University of California, San Diego, had reached the startling conclusion that energy break-even could be achieved with as little as a thousand joules of laser energy, which was not much more than the energy output projected for lasers then under construction at several laboratories. Indeed, Wood soon found that, in theory, break-even was possible at very low energies if the pellet was highly compressed, and Nuckolls, on the basis of his calculations from the early 1960s, concluded that, with optimal pulse shaping, a laser energy of one thousand joules could compress the pellet to meet Wood's conditions. With a new recruit named George Zimmerman, and later with Albert Thiessen, they set out to confirm these theoretical estimates on the computer.

This work, revised and extended, would first be made public in a classic paper by Nuckolls, Wood, Thiessen, and Zimmerman published in the journal *Nature* in 1972.

At about this time I was serving on a committee with Wood and Nuckolls, through which I learned about their amazing findings. I attended an informal meeting of Nuckolls and Wood with the laboratory director, and I was as eager as they were to convince him that Livermore should accelerate its laser fusion research, with funding much greater than the few million dollars per year then being spent. This was 1969, not long after the 1968 meeting at Novosibirsk at which the Soviets presented the results from the T-3 tokamak that would change the course of history in magnetic fusion. But at that time, there was still no certainty that magnetic fusion would ever work in a device of reasonable size. The possibility that laser fusion might work on a modest scale was very appealing to me. While I was busy with my own, unrelated work and had no personal knowledge of weapons design principles or lasers, I continued where I could to encourage others. I recall an open-air ride on a three-wheeled electric cart down the main laboratory thoroughfare, trying to convince a conservative member of the existing laser research team. "If I were you," I said, "I would spend forty days and forty nights in the desert until I knew this thing would work, and go for it."

As enthusiasm for laser fusion spread, the laboratory's director, Michael May, appointed a management group that included the veteran project leader Carl Haussmann. While in 1970 I participated briefly in the search for a laser expert to lead the effort, soon Haussmann began to put together the ingredients of a major program, including the proposal that became the Shiva laser, and in 1972 he recruited John Emmett, then leader of the Naval Research Laboratory's laser program, to head the Livermore effort. After this, I had no involvement except as an admiring observer as the Livermore laser program expanded through a series of devices with exotic names such as Janus, Argus, and Shiva, and finally the giant Nova, completed in December 1984, which is still the workhorse of Livermore laser research, pending construction of the National Ignition Facility (NIF), discussed in chapter 13. During the 1970s, improved calculations led to the conclusion that several hundred thousand joules of laser energy might be needed just to reach break-even, casting doubt on the early optimism. Wisely, Emmett and Haussmann had prepared the way for this eventuality by embarking on a program to develop ever more powerful lasers, which has culminated in the NIF, designed to produce nearly 2 million joules of energy.

Meanwhile, back in 1969 a company named KMS Industries entered the field when the founder, Keeve M. Siegel, learned about laser fusion from Keith Brueckner and started a new company to pursue Brueckner's ideas. I had known Brueckner when I was a graduate student and was aware of his interest in laser fusion, again from casual encounters at scientific gatherings. But I first learned about KMS Fusion when Siegel went public with claims that his new company would demonstrate the feasibility of laser fusion in eighteen months. A true believer, Siegel eventually invested $25 million of company funds and enormous personal energy before he died of a stroke in 1975 while attending congressional hearings to defend his efforts.

While Siegel's interest focused on the use of laser fusion as a commercial electric power source, the U.S. government's funding of laser fusion at Livermore and elsewhere has been provided by the nuclear weapons program in the Department of Energy, which sees laser fusion as a potential small-scale simulator of the type of radiation produced by hydrogen bombs and as a supplement to information heretofore provided by underground tests of nuclear weapons. With the end of the cold war and the current cessation of underground testing, a continuing motivation to fund laser fusion for defense purposes now provides the main support for the NIF. Laser fusion funding in the United States is part of a broader Inertial Confinement Fusion (ICF) program that also includes work at the Sandia National Laboratory on the use of "light ion" beams (for example, lithium ions), as a substitute for laser beams to ignite the pellet. Here "inertial confinement" refers to the fact that the fuel is confined only by the inertia due to the mass of the pellet, during the brief time required for the hot pellet to fly apart.

Despite the continuing military context of ICF research in the United States, Russia, and France, there now exists a worldwide ICF community drawn together by a common interest in the eventual application of this research to the production of electric power. The initial step, to prove that a DT pellet can be ignited to produce a high energy yield, is the same both for energy production and for defense purposes. Impressive progress has been made during the last twenty years, as summarized in chapter 12 and more fully in a recent publication issued in 1995 under auspices of the International Atomic Energy Agency in Vienna.

To understand and appreciate the great progress that has been made in ICF research, let us begin, as we did for the tokamak, by learning how to determine the parameters for an ICF device—specifically, the NIF. This, too, is an exercise I often assigned to my students at Berkeley as a way to introduce them to the concept. Fortunately, we are now free to discuss the ICF concept in greater detail than I was able to do in my classes, thanks to the December 1993 action by the U.S. Department of Energy to remove essentially all secrecy restrictions concerning ICF experiments and the physical principles on which ICF is based. Some of what I am about to discuss was not revealed to American physicists until November 1994, when it was made public at the same meeting of the plasma division of the American Physical Society at which the breakthrough achievement in the TFTR tokamak, the generation of ten megawatts of fusion power, was presented.

The basic idea is sketched in the upper part of figure 5. In an inertial fusion device as in a tokamak, there are three main components, but these are components of a very different nature. They are the DT fuel pellet, or "target," typically only millimeters in diameter; a spherical target chamber to contain the energy release; and a laser or other "driver" (e.g., an array of ion beams) to ignite the pellet. The driver would usually consist of many beams, configured in various ways and fired simultaneously.

Thus, in an inertial confinement device the tiny ICF pellet corresponds to the tokamak plasma, and the laser corresponds to the neutral beams or other means of heating the tokamak plasma to ignition temperatures. But whereas in principle the tokamak plasma can continue to burn by itself once ignited, in an ICF power reactor a new pellet must be injected several times per second, and the laser must be fired repeatedly to ignite and burn each new pellet. The NIF, intended only to demonstrate ignition and modest energy gain, will fire one pellet at a time. The ICF target chamber replaces the vacuum vessel of the tokamak. It, too, has had the air evacuated from it, though for ICF there is no need to maintain the extreme vacuum conditions required for the tokamak. In a power reactor, the inside of the chamber wall would be covered by a thick "blanket" similar to that in a magnetic fusion reactor, but the NIF will not have such a blanket. Overall, the target chamber and blanket would be simpler in an ICF reactor, since there are no large magnets to complicate the design. The complexity lies instead in the powerful laser (or another driver), and in the tiny pellet itself.

As in the tokamak design problem, to make progress my engineering students must again focus on the goal: in this case, to ignite a pellet in the NIF. The formula for success is essentially the same as in the tokamak, namely, the achievement of the Lawson criterion. Again, to join in the game, the reader should recall that the Lawson criterion involves just two numbers, the plasma pressure and the energy confinement time. Whereas for a tokamak the energy confinement time was hard to estimate, for ICF the energy confinement time is just the time it takes for the pellet to fly apart, roughly its radius divided by the speed of the exploding ions (around 100 million centimeters per second after the pellet is ignited by the laser). But what is the plasma pressure in an ICF reaction?

To get started, I suggest that my students first consider a droplet of DT liquid, as Nuckolls and his colleagues did in their landmark paper back in 1972. Hydrogen, and hence DT gas, eventually liquefies if cold enough, at a temperature of minus 253 degrees Celsius, and it actually freezes solid at minus 259 degrees. This may sound awfully cold, but actually such temperatures can easily be achieved by the same method used in your kitchen refrigerator: by repeatedly expanding gases that cool as they expand. In this way, one can create liquid DT that forms spherical droplets, like water dripping from a faucet; with good aim and good timing we could zap one of these droplets with a laser beam as it falls. The density of the droplet, about 0.2 grams per cubic centimeter, is more than 100 million times that of the rarefied DT plasma in the ITER tokamak, and hence at fusion temperatures the pressure would be 100 million times as great also (which is why the pellet flies apart while magnets can confine the pressure in the tokamak).

Recall that it does not matter whether we achieve the Lawson criterion by having a long confinement time (about a second for the tokamak) or a very high pressure, like that in the heated droplet. With a pressure 100 million times that in the tokamak, a large-size droplet one centimeter in radius, flying apart at 100 million centimeters per second, would ignite as well as plasma in the tokamak. So, if a laser could heat the droplet to fusion temperatures, we would have the ingredients for a fairly simple fusion reactor. If all the fuel in the droplet burned, the fusion energy yield would be a thousand times the energy required to heat the fuel, giving a respectable hundredfold overall gain if the laser were 10 percent efficient. Actually, using the minimum Lawson number for ignition, the droplet would fly apart so quickly that only a tiny fraction would burn (which would not be a concern in magnetic fusion reactors, which burn steadily with a continuous supply of fuel). But a somewhat larger droplet with a correspondingly longer confinement time and a larger Lawson number would achieve high gain, and the smaller, one-centimeter droplet would be adequate to demonstrate the onset of ignition, which is the goal of the NIF. To complete our NIF design, we would simply decide to use one-centimeter droplets as targets and design the laser large enough to heat such droplets to a temperature of 100 million degrees Celsius.

Unfortunately, there is a catch. As noted earlier, in his original 1958 proposal that energy could be produced from repeated fusion microexplosions, Nuckolls recognized that the "explosions" had to be scaled down greatly in order to contain the blast and to minimize the size of the "driver" required to ignite the DT fuel. The DT liquid droplet one centimeter in radius described above would have a mass of about one gram, only two thousandths of a pound. But heating one gram of DT to fusion temperatures would require a billion joules, much more than a laser can produce today, and the resulting explosion could be enormous. Fusion reactions are so powerful that the largest fuel quantity one can burn at one time in a fusion reactor is not one gram but only a few milligrams, corresponding to a droplet only a millimeter or so in radius, a speck so tiny that one might not notice it. Even so, a few such tiny droplets, fully burned, would be equivalent to a barrel of oil. However, without the concepts Nuckolls had learned from weapons design, such a tiny pellet would fly apart before it could ignite and burn.

The kind of calculation we have been discussing is what Dawson had in mind when he published his first paper on laser heating of plasmas back in 1964. "I had heard through a friend that Basov was claiming laser heating of a pellet could reach fusion temperatures," he told me recently, "and since I had also been looking at laser heating, I tried the calculation myself." Dismayed by the enormous energy release at ignition, Dawson had suggested in his paper that smaller blasts be magnetically confined. But not having access to nuclear weapons design concepts, he did not realize until later what Nuckolls and his colleagues knew from the outset: that the only way to ignite a few milligrams of DT fuel was to squeeze it first, in order to achieve densities much higher than those of normal liquids or solids. Though a tiny spherical droplet or pellet made tinier by squeezing would fly apart even faster, when the Lawson number is calculated the greater pressure at the higher density obtained by squeezing more than compensates for the reduced confinement time.

With this hint, my engineering students can begin to see how ICF microexplosions could be scaled down to reasonable size. To get a feel for it, I suggest that they ask what would happen if they could somehow compress a tiny droplet to a density a thousand times normal. If a pellet one centimeter in radius meets the Lawson criterion for ignition at normal density, a compressed droplet or pellet only one-thousandth this size would do so at the higher density (since at this density the pressure at ignition temperatures, needed to calculate the Lawson number, is also a thousand times greater).

This tiny compressed pellet, only one-thousandth of a centimeter in radius but a thousand times normal density, would have only one-millionth the mass of the one-centimeter droplet—a mere microgram—and the laser energy required to heat it (a billion joules per gram) would be only one thousand joules, not unlike the early optimistic results at Livermore. With such results to motivate them, I know the students are now ready to think seriously about how laser energy could be applied to squeeze and then heat a pellet, and with what efficiency.

It was how to squeeze, or compress, the fusion fuel that Teller discovered back in 1951, an idea that became second nature to experienced Livermore weapons designers such as Nuckolls. Squeezing liquefied DT, or any other liquid, is outside ordinary experience. It is precisely because liquids cannot be compressed that the hydraulic brakes on your automobile work. When you push on the brake pedal, the incompressible brake fluid transmits this pressure, undiminished, to the brake shoes, which stop the car. Or so it seems. Actually, the liquid only seems to be incompressible because each atom effectively behaves like a tiny balloon filled with high-pressure air. The equivalent pressure in a normal atom, related to what is called its "Fermi energy," is thousands of times the pressure you can apply to the brake pedal. So, if you could muster the strength to push thousands of times harder, the brake fluid would begin to be compressed and the brakes might seem mushy (until the brake tubing exploded). And if you squeeze hard enough on a tiny DT liquid droplet, evenly from all sides, you could squeeze a droplet a few millimeters across to one-tenth that size. This is how ICF works, but it takes enormous pressures to do it.

The method for squeezing an ICF pellet by shining laser beams directly on it is shown in the lower part of figure 5. Squeezing, or compressing, the pellet is accomplished by heating and vaporizing its surface to form a hot plasma atmosphere exploding from the surface. Boiling off material in this way is called "ablation," and the shell of material sacrificed during ablation is called the "ablator." The exploding ablator plasma pushes back on the surface, just as the exhaust ejected from a rocket engine pushes back on the rocket to move it forward. Indeed, the ablation phase is very much like a rocket launch. It is during the launch that the rocket is accelerated to the speed that carries it into orbit. Similarly, ablation serves mainly to get the pellet fuel moving. Because the pellet is a sphere, the outward explosion of the ablator causes the fuel to move inward, to implode, thereby shrinking the radius and compressing the fuel to a smaller volume and higher density.

Ideally, compression continues until the energy of motion of the inwardly imploding fuel atoms, which is proportional to the square of their velocity, equals their Fermi energy in the compressed state. Thus the higher the implosion velocity, the greater the compression. With sufficiently high implosion velocities, the thousandfold compression assumed in our earlier exercise — or even a ten-thousandfold compression — is indeed possible in principle. There is, however, a price to pay. Typically only 10 percent or so of the laser energy contributes to the implosion velocity, the rest being carried away by the exploding ablator. Readers familiar with rockets will notice that this is a poor efficiency compared to that of space vehicles, the reason being that the exploding ablator plasma continues to be heated by the laser as it expands, while a rocket's exhaust cools as it expands. Nonetheless, given the huge reduction in energy requirements achieved by producing a microexplosion with compression, ablative compression is a tremendous bargain for ICF.

Of course, to ignite a pellet, we must heat the pellet as well as compress it. As it turns out, it takes much less energy to compress the pellet even ten-thousandfold than it would take to heat it to the 100 million degrees Celsius required for ignition. But it is not necessary that the laser heat the entire pellet in order to ignite it. This is because of a process called "propagating burn," which is conceptually similar to setting off a firecracker by lighting its fuse. One lights only the fuse (that is, one heats the fuse to ignition), the burning fuse then ignites the fuel around it, this in turn ignites more fuel, and so on, until all the fuel is burning. In ICF research, the "fuse" is called the "hot spot."

Compression itself can create the hot spot, much as compression heats the fuel in a diesel engine. The hot spot would form as a small, hot region in the center of the pellet. The trick is to compress the pellet so as to heat only the hot spot, so that most of the implosion energy is available to overcome the pressure exerted by the Fermi energy of the larger volume of cold fuel surrounding the hot spot. Badly managed, strong shock waves can occur as a result of the implosion, similar to the vibrations that rattle your house if you slam the garage door. Such shock waves, penetrating deeply into the pellet, can heat too much fuel too soon and make further compression difficult. But this problem can be avoided by building up the laser power over time in a prescribed way, the result being an implosion that uses the minimum energy to compress the fuel while also heating and igniting the hot spot just as the maximum density is reached. Matching the laser and pellet design to accomplish these ends is what we mean by pulse shaping. It is the flexibility that laser technology offers to shape the power pulse as we please that is one of the features of laser fusion that so excited Nuckolls and his colleagues in the beginning.

By 1972, when their landmark paper was published in *Nature,* Nuckolls and his colleagues at Livermore had undertaken computer calculations of the processes I have been describing in order to understand pellet implosions in detail. Included in their paper were calculations of the energy gain for a range of laser energies and degrees of compression. The main features can be understood from the simple physical models discussed in the Mathematical Appendix. We will not continue our NIF design exercise here, but I refer the interested reader to that appendix, where the necessary formulas are given, with guidance on how to use them.

Briefly, as we found earlier, the minimum energy required to create a hot spot is greatly reduced with strong compression, while achieving a high gain requires additional energy to assemble a sufficient quantity of compressed fuel around the hot spot. Thus, taking into account the inefficiency of ablative compression, a high gain—say, a hundredfold gain—was found to require about a million joules of laser energy, while the minimum laser energy to create a hot spot and demonstrate ignition ranged from a million joules at low compression to a mere thousand joules if ten-thousandfold compression of the fuel volume could be achieved.

These results, published in 1972, are not very different from similar calculations made today, and the LASNEX computer program that produced these results, modified and improved, continues to be the main tool for designing new ICF experiments at Livermore, including the NIF. These calculations assumed an ideal situation in which the DT droplet remains spherical in shape as it implodes, and pulse shaping is optimized to produce efficient absorption of the laser energy and minimal heating during compression except in the hot spot. Departures from this ideal increase the laser energy required for ignition. The increased understanding of these non-ideal effects that scientists have gained as theory and experimental data have improved over the years now leads them to conclude that a million joules would be needed to demonstrate the ignition of spherical pellets with an energy gain approaching ten. To leave a margin for remaining uncertainties, the NIF is being designed for 1.8 million joules.

Two areas of continuing concern already identified in the 1972 paper were the well-known Rayleigh-Taylor instability, which can distort the pellet shape and mix cold fuel into the hot spot, and the necessity to make the laser irradiation very uniform over the surface of the pellet in order to preserve spherical symmetry during the implosion. The paper also pointed out the possibility of inefficient absorption of laser energy by the pellet and the danger of energetic "suprathermal" (excessively hot) electrons, produced by laser-plasma interactions, which could prematurely heat the fuel during compression. We will meet all of these problems again in chapter 12.

By 1975, as more powerful lasers became available, growing evidence of hot electrons, beam nonuniformity, and incomplete absorption of the laser light led the scientists at Livermore to turn to a different approach. By this time the U.S. government had removed secrecy restrictions on the concepts we have been discussing when applied to the direct irradiation of pellets by laser beams as pictured in figure 5. Others, as we shall see, would continue to explore the "direct drive" approach to ICF and go on to solve the problem of beam nonuniformity. But by 1975 ICF scientists at Livermore and Los Alamos began to concentrate on the still-secret "indirect drive" approach already studied by Nuckolls back in the 1960s, yet another legacy of Teller's ideas from 1951.

For indirect drive, the pellet is mounted inside a tiny can, or "cavity," made of gold, usually referred to by the German name "hohlraum." Even for a device the size of the NIF, the diameter of the can would be only a centimeter or so, and that of the pellet a few millimeters. The laser beams, which enter the cavity through small holes at each end, heat the metal wall so intensely that a metallic plasma is created that is hot enough to emit x-rays. Ideally, the radiation in a hohlraum fills the cavity uniformly as x-rays emitted from one hot region of the surface heat other regions until finally the entire surface is emitting x-rays. In indirect drive, it is the x-ray radiation rather than the laser that heats the pellet, uniformly from all sides if the hohlraum is uniformly hot, thus solving the beam uniformity problem. In practice, x-ray leakage out of the holes where the laser beams enter the cavity and concentration of laser heating where laser beams strike the walls introduce nonuniformity that must be minimized by optimal aiming of the laser beams. An example of an arrangement for doing this is shown in figure 6, which is an actual x-ray photograph of a hohlraum, greatly magnified, showing hot zones where the laser beams strike the wall of the cavity.

Besides solving the beam uniformity problem and symmetrizing the implosion, indirect drive was also thought to offer advantages regarding other problem areas mentioned above. New calculations in the mid-1970s indicated that Rayleigh-Taylor instability was worse than previously expected for direct drive but less severe for indirect drive. A high absorption efficiency was expected, and it was also thought that laser-plasma interactions would be less likely in the metallic plasma created at the surface of the hohlraum, and that in any case only a fraction of any hot electrons produced by laser-plasma interactions at the surface would actually hit the tiny pellet. The two-step process of first converting energy to x-rays and then coupling the x-rays to the pellet does add inefficiencies, but these are partly offset by an improved ablation efficiency for x-ray-driven implosions.

The first experiments on indirect drive at Livermore began in 1976 in the hundred-joule laser facility named Cyclops; these were soon followed by experiments in the two-beam Argus, at one to two thousand joules; and, by the end of 1978, experiments were being run in the ten-thousand-joule Shiva. It was an exciting time, with new challenges at every turn. Throughout this learning process, the workhorse LASNEX computer program helped lead the way, as it still does today. In the tradition of weapons design, Livermore scientists used computer calculations and theoretical models to identify pellet design concepts that ought to achieve ignition and high gain. No mere "droplet," these sophisticated target designs usually included a shell of solid or liquid DT fuel, filled with DT gas and surrounded by one or more shells of plastic or metal utilized in the ablative implosion. The success of these targets would depend on accurate modeling of a wide range of phenomena, including optimal pulse shaping, laser light propagation and absorption, conversion of light to x-rays, radiation and electron transport, implosion symmetry, Rayleigh-Taylor instability, and thermonuclear ignition and burn physics. For each particular target type, these phenomena would be tested in appropriately scaled experiments as John Emmett and his laser team forged ahead to build ever larger lasers to perform these experiments.

Fortunately, though to date no laser experiment exists that is comparable to TFTR in magnetic fusion, experiments combining all aspects of ICF physics have been carried out in a joint effort by Livermore and Los Alamos at the nuclear weapons test site in Nevada, using energy produced by an underground nuclear explosion to drive ICF capsules. These tests, named Halite/Centurion, were intended to study the physics of ICF capsules at high energy long before this will be possible using lasers in the laboratory. While details are not available, in 1990 a committee of the U.S. National Academy of Sciences that reviewed the data concluded: "Since [1986], an outstanding interlaboratory co-operative effort has successfully performed some complex Halite/Centurion experiments that have provided extremely important data. Because of this success, the committee now believes that uncertainties in ignition arise only from mix, symmetry, and laser-plasma interaction, phenomena that can be studied best in laboratory experiments."

Let us now turn to the remarkable progress of these ICF laboratory experiments, through trials and triumphs, paving the way for the NIF.

12 : Catching Up Fast

The first clear evidence in the United States that lasers could heat a pellet to high temperatures was obtained by KMS Fusion in May 1974 and was quickly repeated in experiments at Livermore. These experiments were carried out with simple pellets consisting of a tiny glass bubble filled with DT gas. Though such pellets are not suitable to compress fuel to high densities, plasma produced by rapid heating of the thin glass shell was effective in heating the DT gas. The Livermore experiment with the laser called Janus showed that ion temperatures approaching 40 million degrees Celsius—hotter than the ions in any tokamak at that time—had been achieved. About 10 million neutrons were produced by fusion reactions at this temperature (this corresponds to an energy output of only thirty millionths of a joule, much less than the laser energy). The evidence for heating in all of these experiments was the production of neutrons from DT fusion reactions; and, as in the case of magnetic confinement fusion, early observations of neutrons resulting from laser fusion experiments in the 1960s proved to be of spurious origin, in this case due to the acceleration of deuterium ions by large electric fields created by laser bombardment of slabs of a polymer containing deuterium. Then, in 1976, the larger Argus II laser at Livermore produced about a billion neutrons, permitting for the first time direct measurements showing unambiguously that the neutrons were characteristic of fusion reactions produced by random thermal motion of the fuel ions. These early tests provided the first evidence that the ICF fusion concept might succeed.

I was much impressed by these accomplishments, and I recall, as an "insider" at Livermore, warning my friends outside that ICF represented serious competition for magnetic fusion. As late as October 1977, speaking at a seminar for utility executives, I quoted my Livermore ICF colleagues as expressing the hope that the new Shiva laser nearing completion would achieve "significant" fusion reactions in 1978 and that the still larger Nova laser, then being designed, would demonstrate "high energy gain," thereby proving the "scientific feasibility" of ICF. As it turned out, other scientific issues beginning to be identified intervened, and another decade would pass before the first demonstration of the high degree of pellet compression necessary to realize such ambitious goals.

The early Livermore lasers produced light beams in the near infrared, with a wavelength around one micron (one-millionth of a meter). As researchers were soon to learn, the color of laser light is very important in determining how well that light is absorbed in ICF experiments, and whether or not the energy is absorbed in a way that generates the energetic suprathermal electrons that spoil compression. Although early analysis by Keith Brueckner, Ray Kidder, William Kruer, and others had predicted that laser beams that were too red (i.e., too long in wavelength) could result in high-energy electron production or premature reflection of laser light, the theory of laser-plasma interaction was not sufficiently advanced to predict how long a wavelength was tolerable. The wavelengths of the early lasers used for ICF ranged from ten microns to one micron, one micron being the shortest-wavelength high-power laser available at the time. As it turned out, all of these early lasers had a wavelength that was too long, emitting light that was too red. So, before continuing our story, I had better stop to explain how lasers work, and why a particular laser produces only a certain color of light.

In ICF, the laser produces beams of extremely high power for a short time, just the billionths of a second required to compress and heat the pellet. Very high power is needed to pour energy into the ablator at a rate that matches the boil-off of ablator material required to compress the plasma. It is also necessary to concentrate all of this immense power onto the tiny pellet, only millimeters in diameter. Thus, what we need is a pencil-sharp beam of enormous brightness and power.

Let us first consider how a glass laser works. The intense laser beam is produced in an "amplifier," usually made from a special kind of glass containing the element neodymium, sketched in the upper part of figure 7. The neodymium atoms first absorb energy produced by so-called flashlamps and then re-emit this energy as light of a particular color, characteristic of neodymium. This color is a precise shade of red, in the near-infrared regime just beyond what the eye can see. Fluorescent lights, which use ultraviolet light to excite fluorescent phosphors, work in a similar way, by producing ultraviolet light characteristic of the mercury vapor in the tube. The special feature of neodymium atoms, which causes them to produce a laser beam rather than the ordinary light emitted by a light bulb, is the fact that, once "excited" by absorbing energy, the neodymium atoms do not emit light right away as mercury atoms do. But, by a process called "stimulated emission," even a weak beam of light of the same red color as neodymium light will cause the excited atom to emit its light, thereby adding to the beam. If a large amount of energy has been stored in the neodymium atoms before the weak "trigger" beam is fired, all of this energy will be added to the trigger beam as it passes through the glass, all power flowing in the original direction of the beam. By passing the beam pulse through a succession of such amplifiers, the weak pulse created by the triggering oscillator can be brought up to the enormous power levels required for ICF.

A number of materials besides neodymium can store energy in the excited state long enough to serve as lasers. Maiman's original laser in 1960 produced red light from chromium atoms in ruby crystals. For ICF purposes, ruby quickly gave way to neodymium glass because of the latter's greater ease of manufacture and its superior thermal properties, which were better suited to handle waste heat at high power. Even better, thermally, might be gigantic gas lasers, a popular early version being the carbon dioxide laser that produces infrared light (at a wavelength of ten microns). Gas lasers can also be more energy efficient, if the gas atoms are excited by collisions with electrons, as occurs in fluorescent lights. The currently popular krypton fluoride (KrF) gas laser has this feature, with a potential efficiency of 5 to 7 percent, whereas glass lasers excited by a flashlamp are unlikely to exceed 1 percent efficiency since these lamps produce nearly white light, and neodymium absorbs only the small red component of this light. As we shall see, replacing flashlamps by a new kind of "laser diode" could increase the efficiency of glass lasers to about 10 percent. However, all other issues soon paled when it became clear that the one feature of lasers that was central to further progress in ICF was the color of the light.

The period 1978–83, during which the issue of color dominated ICF progress, has interesting parallels with the period 1965–70 in magnetic fusion (see chapter 3). In both cases, the outcome had a profound effect on the course of events, with the tokamak dominating magnetic fusion and glass lasers dominating ICF. In both cases, advocates from two outstanding scientific institutions—Livermore and Los Alamos, in the case of ICF—held strong views. (Livermore defended glass lasers and Los Alamos its carbon dioxide gas laser.) And in both cases, the culprit was excessively hot electrons.

Researchers understood very well that hot, suprathermal electrons penetrating deep into the pellet were a bad thing, and the possibility that red or infrared light might create suprathermal electrons had been recognized in the paper in *Nature* by Nuckolls and his colleagues back in 1972. Nonetheless, in 1978 the first experiments with the new Shiva laser, which produced ten thousand joules of near-infrared light, were very discouraging, not unlike the experience with the stellarator at Princeton back in 1965. This time it did not take an airlift to Moscow to convince scientists what was happening. "We were awash with hard x-rays," recalls John Holzrichter, who led many of the laser and target experiments at that time. These high-energy x-rays were an unmistakable symptom of suprathermal electrons. The carbon dioxide laser experiments at Los Alamos were only worse. It was a dark time for ICF.

Yet, there were already clues suggesting a solution to the problem, and by 1983 the ICF program was back on track, along the course still followed today. In short order, during 1979 John Lindl at Livermore had identified the problem in Shiva, hypothesized a solution, and conducted experiments that would prove him right. Along with William Kruer, mentioned earlier, Lindl was among the plasma physicists recruited to the Livermore ICF program in the early 1970s in anticipation of the important role that laser-plasma interactions would play in ICF research. Both Kruer and Lindl had received first-class training in plasma physics as students of John Dawson at Princeton. The clue that Lindl recognized was a correlation between the existence of hot electrons and the buildup of plasma filling the hohlraum—to his prepared mind a sure sign that the density had reached the level required for the onset of stimulated Raman scattering, a particularly dangerous laser-plasma instability that we will discuss shortly. The cure, Lindl knew, as a result of years of work by ICF scientists at Livermore and around the world, was a shift from red light to green or ultraviolet light. Confirmed by a series of "scaling" experiments on Shiva that involved variously sized hohlraums and various laser energies and always produced hot electrons when predicted, Lindl's analysis convinced decision makers that the still larger Nova laser, then in the planning stages, should indeed switch from the near-infrared light of its predecessors to green and ultraviolet light.

The breakthrough in laser technology that made it possible to produce ultraviolet light in the Nova laser could not have happened at a better time. The motivation toward the use of a shorter wavelength had been growing throughout the 1970s as experiments in several laboratories began to demonstrate improved absorption of the laser beam in direct-drive experiments using green light, as long anticipated by the theorists. Particularly exciting was the experiment by Edouard Fabre and others at the Ecole Polytechnique in France, reported at a meeting in Boston in late 1979, in which almost complete absorption was achieved using ultraviolet light at a wavelength one-fourth that of neodymium glass lasers. However, these experiments employed relatively small lasers, and no one knew how to build a high-power laser producing ultraviolet light. The KrF gas laser does produce ultraviolet light (at a wavelength of 0.25 microns), but this technology was still far behind that of glass lasers.

It had long been known that the near-infrared light from neodymium glass lasers could be converted to green or even ultraviolet light by "frequency converter" techniques now widely used, but the efficiency of conversion was deemed too low to be useful in high-power lasers. In 1979 Moshe Lubin, director and founder of the Laboratory for Laser Energetics at the University of Rochester, and theory group leader Robert McCrory challenged their scientists to find a method for converting neodymium light to ultraviolet with high efficiency and to carry out tests on the one-beam neodymium laser then under construction as a prototype for a twenty-four-beam system called Omega. With astounding speed, Stephen Craxton, Wolf Seka, and other Rochester scientists did the impossible, and by April 1980 they had demonstrated the conversion of neodymium light to ultraviolet light at a wavelength of 0.35 microns with an efficiency of 80 percent. On the basis of this achievement, McCrory—who would become director at Rochester in 1983—successfully championed the need for ultraviolet light for direct-drive experiments on the Omega ICF facility, and later on Omega Upgrade, just as Lindl had done for indirect-drive experiments on Nova.

In 1980–81, using the new techniques developed at Rochester, Michael Campbell at Livermore carried out the crucial experiments on the powerful Argus laser showing that green and ultraviolet light cured the suprathermal electron problem that had plagued indirect drive experiments on Shiva. As a result of these and its own studies, Los Alamos would soon abandon its far-infrared carbon dioxide laser in favor of KrF lasers and finally curtail its own laser research; and both Livermore and Rochester in the United States and ICF programs in other countries would henceforth convert their neodymium glass lasers to produce ultraviolet light (at 0.35 micron). Research with KrF lasers continues at the Naval Research Laboratory, under the leadership of Stephen Bodner.

Though researchers had appreciated the possibility that plasma effects would spoil things even in hohlraums, actually confronting the disappointing results from indirect drive experiments on Shiva came as something of a shock. The target designers would continue to rely upon and improve LASNEX and other implosion-modeling computer programs, but with a new respect for laser-plasma interactions that must themselves be modeled with ever-increasing sophistication, and a reliance upon ever-improving diagnostic techniques to measure their effects in experiments. A stronger partnership would be forged among the target designers, the laser builders, and the plasma theorists and experimentalists. A stronger partnership would emerge, also, between scientists at the weapons laboratories and scientists at the universities, the University of Rochester having pioneered the technology for converting red light to ultraviolet at high efficiency.

In many respects, the issue of color is to ICF what fluctuations and free energy have been to magnetic fusion. Each involves complex, multiple plasma phenomena that tax to the limits the ability of theorists and experimenters to identify them and to duplicate laboratory results in computer simulations. Again, in ICF as in magnetic fusion, the theoretical underpinnings are Maxwell's equations and the Vlasov equation (see chapter 6), and particle-in-cell computer simulations (see chapter 7). For ICF, researchers must use these computer simulations to calculate the interaction of intense light beams either with the plasma created by ablation (for direct drive ICF) or with plasma created in the hohlraum (for indirect drive ICF).

Let us first consider direct drive. As we learned, the ablation process creates an exploding plasma atmosphere that surrounds the imploding pellet. This plasma atmosphere, called the "corona," absorbs light; whatever light is not absorbed is reflected long before it reaches the surface of the ablator. Ideally, the actual heating of the ablator surface is due to the fact that the laser heats the corona and the corona conducts heat inward to the ablator surface to keep the process going. To understand the color problem for direct drive, let us first understand this ideal situation and then ask what can go wrong.

When plasmas reflect light they do so for the same reasons that an ordinary mirror does. Reflection occurs because free electrons in the plasma, or in the silver backing of the mirror, absorb light energy and collectively re-emit all of this light in the opposite direction. Electrons absorb light energy because, according to Maxwell, light is just a traveling wave of oscillating electric and magnetic fields that accelerate the electrons and therefore give up energy to them. Because the electric field of the light beam is oscillating in time, electrons are accelerated first in one direction and then the other, like the electron current in a radio antenna. This is why they re-radiate the light. If the light beam penetrates to a depth at which the light frequency just matches some natural frequency of oscillation of electrons in the plasma—

for example, the so-called plasma frequency, which increases with density—absorption and re-radiation become so intense that all of the light is reflected. Thus, as the light beam penetrates deeper into the coronal plasma, it would be fully reflected at the density at which the plasma frequency matches the light frequency. This is called the "critical density."

Fortunately, some of the energy of the oscillating electrons is converted to heat when the electrons collide with each other or with plasma ions, thereby creating thermal motion (heat) that is no longer synchronous with the light. Thus the light wave heats the electrons through collisions, the same kind of binary collision process we encountered in our discussion of entropy in chapter 6. The actual heating of the ablator is caused by the conduction of this electron heat through the corona until it reaches the dense ablator surface, as already noted. This useful thermal conduction of heat by electrons, due to the collision of hotter electrons with neighboring colder electrons, is not to be confused with the presence of the dangerous suprathermal electrons, to which we now turn our attention.

In addition to the useful absorption of laser energy discussed above, there are also collective plasma phenomena, often more important, that may either reflect or absorb the laser beam. One such process is the so-called resonance absorption that occurs in the vicinity of the critical density at which ordinary reflection occurs. This resonance absorption competes with collisional absorption. Other collective processes can cause reflection to occur prematurely at lower plasma densities farther out in the corona. These processes, in which the laser beam excites plasma waves, include stimulated Brillouin scattering (SBS), and stimulated Raman scattering (SRS). Unfortunately, though both resonance absorption and SRS do absorb energy from the laser beam (a good thing), they largely do so by accelerating electrons to high energy—that is, by creating the dangerous suprathermal electrons that penetrate clear through the corona, finally stopping deep in the compressed fuel.

Which collective processes are important depends on the situation. For direct drive ICF, all three of the processes mentioned above have been important at one time or another, in the form of premature reflection by SBS so that light is not absorbed, or as resonance absorption and SRS that efficiently absorb the light but convert part of the energy to suprathermal electrons (by the Landau damping of plasma waves, discussed in chapter 6). For the indirect drive experiments on Shiva, it was SRS that Lindl had identified as the source of suprathermal electrons, the clue being that SRS sets in very strongly when the plasma density in the hohlraum reaches levels between one-tenth and one-fourth the critical density at which ordinary reflection would occur. For indirect drive experiments on the NIF, SBS may also become important, forcing the density threshold even lower, to about one-tenth of the critical density, to avoid SBS and SRS effects that could deflect the laser beams off course and spoil the aiming needed to maintain radiation symmetry in the hohlraum.

Finally, let us try to understand how collisional absorption is enhanced and suprathermal electrons are suppressed by switching from red light to green or ultraviolet light. That this occurs has been clearly demonstrated in experiments showing that absorption decreases as the beam intensity is increased for almost all wavelengths but that this phenomenon is markedly less pronounced for ultraviolet light at wavelengths of 0.35 microns and 0.25 microns. For wavelengths of 1.0 micron or less, this can be understood, qualitatively, by considering only collisional absorption. As we learned in chapter 7, the collision rate goes down as the temperature increases. This accounts for the drop-off in absorption observed at higher beam intensity, since a more powerful beam heats the coronal plasma to a higher temperature. However, the collision rate increases and absorption improves if the absorption takes place at a higher density, which accounts for the observed better absorption of green or ultraviolet light at any given beam intensity. Green and ultraviolet light have higher frequencies than does red light. Since reflection occurs where the light frequency matches the plasma frequency, which increases with density, the higher-frequency green or ultraviolet light penetrates to a greater depth in the corona, where the density is higher (again, unless it is prematurely reflected by the collective SBS or SRS processes). In this respect, green is better than red, blue is better than green, violet and ultraviolet are better than blue, and so on.

In principle, the collective processes can be avoided if the beam intensity falls below a threshold at which the rate of excitation of the plasma waves is less than the rate at which these waves are damped by other processes. These thresholds can be calculated theoretically with good accuracy, the threshold for trouble being higher at shorter wavelengths. This gave the Livermore scientists the confidence to predict that the problems encountered on Shiva with near-infrared light would be solved with ultraviolet light in Nova.

In practice, the desire to get everything possible out of the equipment drives the experimenter to operate as far above these thresholds as possible, until premature reflection or the occurrence of suprathermal electrons stops progress. Hence, theorists also try to push beyond the threshold for the SBS and SRS processes to calculate the consequences of these collective processes, using the PIC computer simulations mentioned above. Thus far, such calculations can accurately indicate trends, but quantitative prediction is beyond their capability. Like the prediction of the energy confinement time for tokamaks, the prediction of efficient absorption of laser light with minimal production of unwanted suprathermal electrons in the NIF is still largely based on experimental results. It is achieving the predicted conditions for efficient absorption without suprathermals, plus improved understanding of the hydrodynamic Rayleigh-Taylor instability discussed below, that leads us to the NIF design, which uses 1.8 million joules in order to reach ignition rather than relying on the more optimistic estimates included in the landmark paper by Nuckolls and his colleagues back in 1972. It now appears that obtaining ignition at energies much less than the NIF uses would require laser beam intensities well above the threshold for the deleterious effects of laser-plasma interactions, even with ultraviolet lasers.

Having resolved the color issue for the time being, the ICF physics community during the last decade has turned its attention primarily to understanding the physics and design issues involved in compressing pellets to high density. As mentioned above, compression physics was already well advanced when the American ICF program was formalized back in 1972, and advanced hydrodynamic computer models for ICF, such as the LASNEX computer program at Livermore, were already being undertaken. However, detailed comparison of computer predictions with laser experiments in the laboratory had to await the development of large lasers producing light at suitable wavelengths, and experimental techniques to measure the appropriate quantities with adequate precision. The largest of these lasers are Nova, at Livermore, producing forty-five thousand joules; Omega, at the University of Rochester, recently upgraded to about the same energy; and the ten-thousand-joule Gekko XII at Osaka, Japan—all with the capability of producing light at the ultraviolet wavelength of 0.35 microns. Los Alamos scientists participate in experiments on Nova.

Like its magnetic fusion counterpart, the ICF program today finds itself within shooting distance of ignition—about a factor of ten away, in terms of the Lawson number for the hot spot. A single indicator such as the Lawson number only tells us where we have been but not how we got there or how to move forward. However, by probing in various directions and comparing the results with theory, scientists can find the path of progress. Always limited by the capability of the lasers available at the time, some experiments have probed toward higher temperatures, others toward higher densities. Using the simple microballoon "exploding pusher" targets first employed at KMS twenty years ago, temperatures approaching 100 million degrees Celsius have now been obtained at the Nova and Omega Upgrade facilities, producing record yields of fusion neutrons, millions more than those first results that so impressed me back in 1974. Using other kinds of targets, in 1988 researchers at Rochester achieved a DT density of twenty or more grams per cubic centimeter, one hundred times the density of liquified DT, and in 1991 the Osaka group compressed a DT-impregnated plastic shell to a density of six hundred grams per cubic centimeter, though in both cases a poor yield of neutrons indicated that strong hydrodynamic instability had precluded achieving a high temperature in the hot spot.

Armed with a wealth of new experimental information and ever more sophisticated computer models, by 1990 Livermore scientists had begun to piece together the requirements for an experiment to bridge the ignition gap, paving the way for the NIF. According to the computer models, carefully optimizing the laser configuration and pellet design to achieve stable compression to higher densities would allow the hot spot temperature and the density to be increased simultaneously. Then a fortyfold increase in laser energy beyond Nova could produce ignition and a millionfold increase in fusion energy, an enormous return on investment.

I first heard this exciting story in May 1990, as a member of a review committee chartered by the U.S. Department of Energy to review both the magnetic and inertial fusion programs and make policy recommendations for developing fusion as an energy option. At that time I had already left Livermore and was somewhat out of touch with ICF progress, though I had begun to relearn ICF fundamentals in order to include them in my fusion courses at Berkeley. By September 1990 our committee, called the Fusion Policy Advisory Committee (FPAC), had made recommendations to transfer the small component of the ICF program devoted to electric power production to the Office of Fusion Energy, which also manages magnetic confinement fusion, in order to promote cooperation between these programs. By then, the ICF and magnetic confinement scientific communities were meeting together regularly at the annual plasma meetings of the American Physical Society, and their exchange of information was soon made easier by a loosening of restrictions on formerly secret ICF information, also recommended by our committee. In recognition of the excellent science being conducted by the ICF program, the American Physical Society had awarded its prestigious Maxwell Prize to John Nuckolls in 1981 for his pioneering role in ICF, and in 1990 it awarded the Maxwell Prize to another Livermore ICF scientist, William Kruer, for his theoretical work on complex laser-plasma interactions. Also, the Department of Energy had presented its Ernest O. Lawrence Award to other ICF pioneers—to John Emmett in 1977 for his role in developing high-power lasers, to George Zimmerman in 1983 for the development of the LASNEX computer program, to Gene McCall at Los Alamos in 1988 for his work on laser-plasma interactions, and to Pace Vandevender at Sandia in 1991 for his work on pulsed power. In 1994 it would present the award jointly to Michael Campbell and John Lindl at Livermore, for their experimental and theoretical work on indirect drive.

A turning point in our FPAC deliberations was a presentation by Lindl describing his computer calculations showing an "ignition cliff" not far beyond what was being accomplished on Nova at Livermore. This "cliff" refers to the conditions at which the ratio of the fusion energy yield to the laser energy required to obtain it increases steeply. At this time, the National Academy of Sciences was conducting a review of ICF, including a long-range proposal for an ICF facility called the Laboratory Microfusion Facility (LMF) that would be comparable to ITER and would require a much larger laser than the NIF. It was partly in response to skepticism about such a large step that Lindl and other scientists at Livermore began thinking about a more modest step, building on existing facilities.

"Since the January 1990 issuance of the National Academy of Sciences interim report," our own committee report stated, "there has been an important theoretical advance at Livermore. Numerical simulations have indicated that by going to shorter pulses and higher hohlraum temperatures, already attained on Nova, it should be possible to achieve ignition and modest (5 to 10) gain implosions with a one to two million joule driver. This could be accomplished by an upgrade of the Nova laser in the existing building. This approach would be considerably (2–3 times) less expensive than a five to ten million joule driver and represents, in our opinion, a more logical next step than a full LMF. Pellet ignition by about the year 2000 is an attractive goal that can probably be attained if construction of the Nova Upgrade starts in 1994, as we recommend."

With some changes to serve the wider national and international ICF communities, this proposal by Livermore to upgrade Nova eventually led to the NIF, which has now been endorsed by several review boards (see chapter 13). In October 1994, Secretary of Energy Hazel O'Leary announced her support for the construction of the NIF, at a site yet to be determined, though Livermore has been named as the preferred site. Funding for the NIF final design was included in the 1996 budget for ICF, and the work is proceeding. Meanwhile, in May 1995, the French government announced plans to build its own version of the NIF, also to produce 1.8-million-joule pulses of 0.35-micron light. The existing French laser fusion facility at Limeil, operational since 1986, was engineered by Livermore and constructed jointly by a Livermore-Limeil team.

The scientific enthusiasm for the NIF, generally shared throughout the ICF community, continues to rest heavily on the predictions of computer calculations and on experimentally derived refinements to increase the validity of the computer simulations. By and large, Lindl's 1990 conclusions have stood the test of time. A new experimental campaign termed "Precision Nova" was launched at Livermore to pursue the directions indicated by these computer calculations, with improved diagnostics and great care to achieve the most uniform target irradiation possible with only ten beams on Nova. In one Precision Nova experiment, DT gas was compressed to 20 grams per cubic centimeter—one hundred times the density of liquified DT—with no significant disruption of the hot spot by hydrodynamic instability. In these experiments, the "main fuel" was actually glass, the target being a glass bubble filled with DT gas and surrounded by a plastic ablator to enhance compression. The glass shell itself, playing the role of the main fuel in a real target, was compressed to a density of 150 grams per cubic centimeter, and the DT gas was simultaneously compressed to 20 grams per cubic centimeter, with a temperature around 10 million degrees Celsius in the hot spot. This experiment combined most elements of the compression scenario that will be played out to achieve ignition on the NIF.

Compression is to ICF what the pressure limit is to magnetic fusion. For the designer, any limitation on what is allowed by the laws of physics is significant. Thus, one of the linchpins of magnetic fusion design is the guideline provided by the agreement between experimental data and Troyon's calculation of the pressure limit (see chapter 2). The corresponding results for ICF, showing the limits on design that must be obeyed to avoid excessive distortion of the pellet during compression, concern the Rayleigh-Taylor instability.

In ICF, a spherical pellet must remain nearly spherical as it is compressed. Any bumpiness of the surface due to the manufacturing process or any nonuniformity in the heating of the ablator spoils this ideal, the more so because small bumps grow rapidly owing to the Rayleigh-Taylor instability mechanism that causes depressions in the surface to grow by pushing material into the already bulging bumps. Therefore, the design restriction due to the Rayleigh-Taylor instability is that the bumps due to manufacturing must be small enough and their rate of growth slow enough that the pellet is still roughly spherical when compression ceases. The outcome is determined by the time the ablator has blown away, leaving an imploding shell of compressed fuel with whatever bumpy imperfections remain from the pellet's manufacture. Knowing the growth rate of these bumps as the implosion proceeds, we could calculate the consequences. As it turns out, bumps grow excessively if the compressed shell is too thin, while a shell that is too thick would be difficult to compress. Hence Rayleigh-Taylor instability ultimately sets the limits on how much the fuel can be compressed. To relate the allowed compression ratios to tolerances on pellet surface roughness, it is necessary to be able to calculate with good accuracy how bumps will grow as a result of Rayleigh-Taylor instability.

Indirect-drive experiments on Nova yield excellent agreement between experimental results and predictions for Rayleigh-Taylor instability. Similarly good agreement has been obtained with direct drive, in work carried out by research teams in England and Japan as well as the United States. To achieve precise agreement with theory and to see the effect of large and small asymmetries, a known degree of bumpiness is introduced from the start. Bumps of various sizes are investigated in order to understand both the "linear" case, in which the bumps are very small, and the "nonlinear" case, in which large bumps begin to affect each other, a greater challenge for the theory. In November 1995, the Division of Plasma Physics of the American Physical Society honored this work by presenting an "Excellence in Plasma Physics" award to a joint team from Livermore and the University of Rochester.

Obtaining such experimental data, in experiments lasting only a billionth of a second, is extremely difficult. In order to ensure that the Nova experiments are carefully controlled, the pellet is replaced by a flat plate initially made bumpy to a prescribed degree. A hole in the hohlraum cavity allows x-rays generated in the hohlraum to reach the plate. These strong x-rays cause ablation that drives the plate forward. An additional, weaker source of x-rays is used to take repeated x-ray images as the bumps grow during acceleration, as they would on the ablator surface of an actual spherical pellet. This image is one-dimensional, taken through a slit. Sweeping the image across a film produces a sequence of slit images in which the growth of bumps over time is detected as a progressive lightening of the image where a bump is located, just as a medical x-ray distinguishes light bones from darker tissue. This technique, called a "streak camera," is frequently used in ICF and other fields of plasma physics to record information—in this case the growing contrast between light bumps and the darker valleys between bumps—as it develops over time. Applying such techniques to ICF, in which information must be processed in a billionth of a second, is particularly challenging. Older versions of the streak camera used a rotating mirror to cause the image to move across a film. For ICF, sweeping is accomplished electronically. The x-rays first produce electrons where they strike a specially coated surface, thereby turning the x-ray image into an equivalent electron image. These electrons in turn strike a different specially prepared surface, similar to a television screen, producing a real picture that can be recorded photographically. Rapidly varying electric fields that deflect the electrons sweep the picture across the screen to produce a continuous sequence of one-dimensional pictures spread out in time.

The accurate prediction of Rayleigh-Taylor instability at the ablator surface and of the limits this instability imposes on design is only one of several pieces of evidence leading to confidence in the capability of ICF computer models to predict compression and heating. Other experiments have been carried out to isolate and calibrate phenomena such as the mixing of ablator material with the fuel due to a bumpy ablator surface, and the growth of surface distortions due to nonuniform heating and acceleration of the ablator surface. Roughly speaking, these studies have shown that no more than a 1 percent variation in heating power can be tolerated, thereby placing strict demands on the design of laser systems both for direct and for indirect drive. Experiments have shown that these demands can be met with indirect drive.

Other experiments have been designed to create and measure the propagation of shock waves, important in determining the optimum rate at which the laser power should be increased, first to sustain compression without heating and then to heat the "hot spot." Finally, integrated computer programs including all phenomena simultaneously have demonstrated gratifying agreement with pellet implosion experiments. Spherically symmetric compression reducing pellet volume by a factor of roughly 100 has been demonstrated with direct drive on the Omega laser at the University of

Rochester, in good agreement with computer simulations up to the point at which hydrodynamic instability finally set in.

Despite all that has been learned about Rayleigh-Taylor instability and implosion hydrodynamics, embodied in the ICF computer models, there is still no clear choice between the direct drive and indirect drive approaches to ICF, and the differences between them have become the focus of friendly rivalries between ICF laboratories. For a long time, because of secrecy restrictions, work on indirect drive in the United States was the special province of Livermore and Los Alamos, while the University of Rochester and other laser laboratories necessarily concentrated on direct drive. In the early days, the symmetrizing effect of the hohlraum seemed to give indirect drive a clear edge, but in the end the wider knowledge obtained from pursuing both approaches has been very beneficial to ICF. On the basis of the work at Rochester, it is probably fair to say that achieving the symmetry required for direct drive has proved to be more feasible than critics feared, while achieving adequate symmetry with indirect drive has proved to be more challenging than advocates first thought. As a result, the NIF will accommodate both direct drive and indirect drive experiments.

Various ICF advisory committees, including the FPAC on which I served in 1990, have seen the wisdom of preserving several approaches at this stage of ICF development. This was the main motivation for our recommendation to proceed with the Omega Upgrade at Rochester, which is configured for direct drive experiments, whereas Nova, at Livermore, is configured primarily for indirect drive experiments. Like Nova, the Omega Upgrade is a powerful neodymium glass laser with frequency conversion to the ultraviolet (a wavelength of 0.35 micron). Now completed, it is to be the direct-drive counterpart to Nova, with sixty beams producing up to forty-five thousand joules of ultraviolet light pulses a billionth of a second in duration. When I was writing this, in August 1995, an announcement had just appeared in *Physics Today* about the completion of the construction of the Omega Upgrade and its initial operation, and also about the completion of the Nike KrF gas laser at the Naval Research Laboratory, also recommended by our FPAC committee. The initial experiments with the Omega Upgrade have been encouraging.

Both the Omega Upgrade and the Nike programs have as important goals the development of "smoothing" techniques that improve the uniformity of laser irradiation of directly driven targets. The technique used in the

Omega Upgrade is designed to overcome imperfections in optical components that distort the light beam, causing it to illuminate the pellet nonuniformly. By a clever combination of optical elements and an "electro-optic modulator" that spreads the frequency, or color, of the light, the single beam is converted to a bundle of "beamlets" the size of the pellet, all overlapping, so that on average a much smoother illumination is obtained. This technique will now be used on the NIF. The Omega Upgrade is also designed to achieve the same energy and power in each beam, with an accuracy sufficient to provide direct drive irradiation that is uniform to 1 percent, thus meeting the experimentally determined criterion for pellet compression.

Though perhaps a decade behind glass lasers, the Nike KrF gas laser offers potential advantages in the long run, in part because smoothing techniques simpler than those now employed on the Omega Upgrade glass lasers can be applied directly to the low-power trigger beam, and because the natural color of the light emitted is in the ultraviolet range (0.25 micron), so that frequency conversion is not required. According to the August 1995 article in *Physics Today*, even a single beam of KrF light produced by the Nike laser is already uniform to within 1 percent, and the forty-four overlapping beams of the complete Nike system should provide target irradiation uniform to within 0.3 percent, even better than can be achieved with glass lasers with smoothing. Advocates of KrF lasers also point to their higher efficiency due to electron beam excitation (see chapter 11) and their ability to handle waste heat at the fast repetition rate required for energy production.

Meanwhile, not to be outdone, scientists at Livermore, in France, in Japan, and elsewhere have continued to develop glass laser technology with the aim of increasing glass lasers' efficiency and repetition rate. It now appears that a combination of gas cooling between shots and a more efficient excitation scheme that reduces the amount of heat wasted might make feasible neodymium glass lasers suitable for power reactors, with an efficiency of about 10 percent firing ten times per second. The technology development that makes this possible is the "laser diode," a low-power, very-long-pulse-length laser that can produce light within a narrow range of wavelengths similar to the wavelength of neodymium light.

We now turn to the design of the NIF, the technologies required for it, and the role of the NIF in developing ICF as a future energy source.

13 : The National Ignition Facility

A walk through the Nova laser building at Livermore is an extraordinary experience. I often took visitors there, and to Shiva before that, when they came to see and discuss our magnetic confinement fusion experiments. The contrast between our very industrial-looking magnets, vacuum vessels, and electrical equipment and the space-age lasers never failed to impress me. One important difference between the laser building and a magnetic confinement facility is the necessity to maintain a higher degree of cleanliness around the laser, to avoid accumulating dust on exposed glass. A visitor must don a lab coat and shoe coverings. Another important difference is the necessity to mount the laser equipment very solidly to avoid the slightest vibration that could cause misalignment of the beams. For example, the Shiva laser had been mounted on an enormous slab of concrete. In the 1980 earthquake mentioned earlier, the main damage to Shiva occurred when the shaking of the earth produced sideways forces violent enough to shear in two the giant bolts that attached the laser to the slab.

Nova consists of ten identical beamlines mounted side by side in a large, open room that has sixty thousand square feet of floor space and is as high as a five-story building. A Nova beamline, which serves as the laser amplifier described in the previous chapter, consists of a series of neodymium glass slabs excited by flashlamps something like a neon sign but filled with xenon gas, not neon. The electrical energy needed to fire the flashlamps, ultimately the source of the laser energy, is first stored in "low-tech" devices not very different from the "capacitor banks" used in magnetic fusion research and many other applications. The burst of light from the flashlamps lasts about five hundred millionths of a second, during which the near-infrared portion is absorbed by "exciting" neodymium atoms in the glass. Then all of this energy serves to amplify a light pulse only a billionth of a second in duration, briefly about a million times the power of the flashlamps.

Because all ten beamlines are the same, technology development for Nova could be carried out for just one beamline. All of the laser development at Livermore has been carried out in this way, starting in 1975 with the

Cyclops one-beam laser and the Argus two-beam laser used to test optical designs for the twenty-beam Shiva.

This modular approach to developing avant-garde, large lasers has served the ICF community well as researchers sought to validate concepts and ferret out unsuspected problems. For example, many experts had predicted that glass lasers would fail at high power due to a "self-focusing" of the beam that would only become evident at high power and that would make it impossible to maintain the perfect focus of the trigger beam as it passed from one amplifier to the next. Indeed, despite the success of its two-beam Janus laser, which was completed in 1974, the Livermore team did encounter self-focusing in the more powerful Cyclops laser and concluded that new technology and a more systematic approach to modeling beam optics were needed to meet the goal of Shiva and the still higher-power lasers that followed it. Self-focusing causes the laser beam to break up into filaments of such intensity that they damage the glass. The new technology, called a "spatial filter," focuses the beam through a tiny hole, something like the hole in a pinhole camera, to "scrape off" the filaments before they grow. Since the little filaments tend to diverge and miss the hole, the emerging beam is much more uniform. Convincing tests of spatial filters on the Argus laser paved the way for implementing them on Shiva. Thanks to these filters, Shiva was successfully producing ten-thousand-joule pulses by 1978.

The prototype beamline for Nova, called Novette, was completed in 1983. As related in chapter 12, by the end of the 1970s experiments on Shiva and elsewhere had shown that the near-infrared light from neodymium was inadequate for ICF experiments. Following the approach already tested at Rochester and on Argus, the Livermore team had designed Novette to convert neodymium light efficiently to green or ultraviolet light.

The "frequency converter" that is used to turn neodymium light into shorter-wavelength light works as follows. The basic idea, long used in musical instruments and in radio and other electronic gear, relies on the fact that natural oscillators often produce many different overtones, or "harmonics," related to each other in a precise way. Green light with a wavelength of 0.53 micron is the second harmonic of neodymium light, this shade of green having a wavelength exactly half that of the 1.06-micron neodymium light. The third harmonic, with one-third the wavelength (0.35 micron), lies in the near ultraviolet, just beyond what the human eye can see.

Certain materials will produce harmonics of bright light passing through them. One such material is a large crystal of potassium dihydrogen phosphate (KDP). When a laser beam is passed through a KDP crystal, the electrons in the crystal oscillate at the frequency of the laser beam. At low intensity, the electron motion is oscillatory at a pure frequency characteristic of the color of the laser light. However, as the laser intensity is increased, this pure oscillation becomes perturbed and develops a second-harmonic component that in turn generates a laser beam at the second harmonic, in the same way that a bugle player can jump an octave from low C to high C. As the fundamental (near-infrared) beam and the second-harmonic (green) beam propagate together, the second-harmonic beam is constantly reinforced and can in principle extract all the energy from the fundamental beam. Similarly, if a fundamental and a second-harmonic beam are passed through a crystal, the third harmonic can be generated. This scheme was used to convert neodymium light to ultraviolet 0.35-micron light in Novette, at an efficiency up to 80 percent. Actually, there are two KDP crystals in series, one that converts part of the red light to green, and a second that mixes the remaining red light and the green light to obtain the third-harmonic ultraviolet. The same scheme will be used in the NIF.

As noted in chapter 12, the breakthrough in producing third-harmonic ultraviolet light efficiently by mixing red and green light occurred during 1979–80 at the University of Rochester, in response to a challenge from the laser laboratory director, Moshe Lubin, and the theory group leader, Robert McCrory. Taking up their challenge, Rochester theorist Stephen Craxton soon provided a key idea. Starting with the two-crystal scheme mentioned above, he traced the inefficiency of earlier efforts to a bad mix of green and red light entering the second crystal, where the reaction producing ultraviolet light by combining red and green occurs. Just as in chemical reactions, the mix must be just right to get the desired result, but maintaining a constant mix of laser colors was impractical as the laser power rose to full intensity during a pulse: in passing through the first KDP crystal a laser beam of rising intensity becomes more and more green, thereby changing the mix. The solution, Craxton concluded, was to spoil the conversion of red to green light in the first crystal just enough to obtain the best mix of red and green light entering the second crystal. His first idea for doing this involved tilting the first crystal by a tiny but precise angle. Within a few weeks, the idea was ready for testing. "The experiments were due to begin on a Monday," he recalled recently, "but on the Friday before that, I had a better idea." On Monday, he shared the new idea, which involved changing the polarization of the red light, with his colleagues, who proceeded to test both ideas in sequence, all in a few days. Both ideas worked as predicted, and in the end, both have proved useful in practice, the polarization scheme having been adopted in the Omega and Nova lasers, and the tilt scheme in the NIF design. Sadly, Lubin did not live to see his vision fulfilled in the new Omega Upgrade at Rochester.

By the time Nova was built, enough had been learned about the optics of high-power beams in glass to increase the energy produced per beamline more than tenfold, whereby 10 beams on Nova could produce up to ten times the energy of the 20-beam Shiva. The NIF makes similar strides relative to Nova, in terms of the increase in laser energy relative to the cost of the facility. The NIF design requires 192 beams to produce the 1.8-megajoule pulses required to reach ignition. The NIF beamlines differ from those of Nova in the fact that, in the NIF beamline, the light passes back and forth through the amplifiers in order to gain a higher amplification than could be obtained in one pass. This compact design approach lowers the cost. Also, thanks to new improvements in glass technology that allow greater intensity without damaging the glass, the final beam "fluence" (energy per unit area) is higher than in Nova.

A prototype NIF beamline, called Beamlet, was completed at Livermore in 1994. The Beamlet has already produced a pulse of ultraviolet light three billionths of a second in duration, with an energy of sixty-four hundred joules. These results meet NIF requirements, when scaled to the somewhat larger aperture planned for the actual beamlines.

The Beamlet components, corresponding to one beamline of the NIF, are shown in the lower part of figure 7. "Spatial filters" like those used on Shiva and Nova serve to smooth out the effects of optical filamenting at high power. Two new components, not present in Nova, are the Pockels cell and the "deformable mirror" that allow the beam to pass back and forth through the amplifiers.

The deformable mirror helps to compensate for distortions in the beam as it travels along the beamline. It does this by reflecting different parts of the beam at different angles so as to bend stray rays back into proper focus. Optical sensors that examine the beam send signals to "actuators" mounted at many locations on the back side of the mirror, which then apply pressure to bend different portions of the flexible mirror surface to just the extent needed to straighten out the beam for subsequent shots. Similar deformable mirrors are being installed on the Lick Observatory telescope and the huge

Keck telescope in Hawaii to produce a clear image of stars that seem to the eye to "twinkle" because of turbulence in the atmosphere.

The large Pockels cell in the NIF beamline serves as a "switch" to direct traffic as the beam passes back and forth, being amplified as it goes. Together with the "polarizer" shown in the lower part of figure 7, the Pockels cell allows us to reuse the amplifiers several times, enabling the beamline to produce as much amplification as a much longer beamline.

To see how this works, let us follow the beam as it travels through the NIF beamline. The trigger beam, or "injection pulse," that begins the process of generating the laser beam enters the main beamline near the point where the beam eventually passes through the "frequency converter" on its way to the target (fig. 7, *below, right*). This is a one-joule pulse of 1.06-micron neodymium light generated by a master oscillator and preamplifier not shown in the figure. This small injection pulse from the preamplifier, which is infrared in color to match neodymium light, is reflected from a small mirror (labeled *LMo*) that directs the beam backwards (i.e., away from the target chamber) through the chain of lenses, mirrors, and neodymium glass amplifiers. Upon reflecting from the deformable mirror *LM1* at the far left, the beam pulse retraces its path and gains additional energy as it passes through the amplifiers a second time. At this point, the Pockels cell is turned on and intervenes to prevent the light from reflecting from the polarizer as it did in its first journey backward along the beamline. Instead, the beam passes right through the polarizer and reflects backward again, at the mirror labeled *LM2*, thereby passing once more through the two "cavity" amplifiers, on to the deformable mirror, and back again. Finally, when the Pockels cell is turned off, the amplified neodymium light beam does reflect from the polarizer to pass through the "booster" amplifier and the frequency converter and on to the target as an intense pulse of ultraviolet light.

To understand how the Pockels cell performs such magic, we must consider how the polarizer itself works. A pure light wave has an orientation, or polarization, transverse to its direction of travel (technically, the oscillating electric field of the wave has this orientation). Certain materials will transmit a pure light wave when they are aligned with its direction of polarization and will reflect the light when aligned at right angles to the direction of polarization. Polaroid sunglasses use two such polarizers, twisted relative to each other, to partially reflect sunlight consisting of many waves with various directions of polarization.

In the NIF beamline, the polarizer is oriented to reflect light polarized in the same direction as the pure beam produced by the master oscillator. Thus, the original trigger beam reflects from the polarizer and continues backwards along the beam path. Before the beam returns, the Pockels cell is energized (all in a few ten-billionths of a second!). When electrically energized, the KDP crystal in the Pockels cell causes the polarization of the beam to twist ninety degrees. Then the beam no longer reflects from the polarizer but instead passes right through it. After reflecting from the mirror labeled *LM2* the beam again passes through the Pockels cell, which twists the beam back to its original polarization, which it retains as it travels once more through the cavity amplifiers. The Pockels cell is then turned off, allowing the returning beam to be reflected by the polarizer and proceed on its way through the booster amplifier and frequency converter, out of the system.

Researchers' confidence that they can implement such optical gymnastics, most of which have now been demonstrated in the Beamlet, has played a major role in reducing the projected cost of the NIF. Not only do these techniques require fewer optical components but the shorter beamlines made possible by multiple passes through the system reduce the size of the building to about twice the size of the Nova building. An artist's drawing of the complete facility, including the target chamber, is shown in figure 8. Roughly the dimensions of a football field, the NIF would fit comfortably inside a modern sports arena such as the Metrodome in Minneapolis. The projected cost, in the range of $1 billion, including development costs, is about what the TFTR or JET tokamak would cost if built today.

Just as important as the laser to ensuring success in the NIF is the design of the target pellet. While pellet design is the ICF "architectural" counterpart of magnetic field design for magnetic fusion, the minuscule size of ICF pellets has given freer rein to inventors, allowing many more options to be explored and discarded. Pellet design is closely coupled to the computer models that allow scientists to investigate and optimize many approaches (see chapter 12). The designs that have been investigated include shells within shells, each made of some different material serving to generate and transmit the implosive force in the best way to achieve the desired density and final heating of the hot spot. Over the years, pellet fabricators have vied with designers, undaunted by extreme demands for precision on a tiny scale, which put to shame the achievements of even those craftsmen most skillful at constructing ships in bottles.

A typical technique involves producing a thin shell to contain the fuel and filling the shell with DT fuel in the desired configuration. Later, other coatings of plastic or metal can be added by exposing the pellet to plastic or metal vapors that deposit on the pellet surface. The shell can be formed in various ways—for example, by forming falling droplets made of plastic dissolved in an organic solvent. Surface tension causes the droplets to assume a spherical shape as they fall, and if the temperature is kept sufficiently high the solvent evaporates, leaving a spherical plastic shell. This technique works well for various combinations of plastic and solvent, for shells up to a few millimeters in diameter.

A breakthrough occurred in the 1970s with regard to so-called cryogenic targets, whose name refers to a method for creating the shell of frozen DT that serves as the main fuel in hollow pellets of the type assumed in our discussion of ablative compression and hot spot formation in chapter 11. The plastic shells whose manufacture is described above can be filled with DT gas in several ways—for example, by placing the shells in DT gas at high pressure and waiting for the DT to diffuse inside. Exposed to very low ("cryogenic") temperatures, the gas freezes to form a solid layer of DT "snow" on the inner surface of the shell. As it happens, heat produced by the slow radioactive decay of the tritium causes thick spots to sublimate (or transform directly from solid to gas) as a snow bank would, and the gas recondenses to fill in "valleys," forming a very smooth fuel layer of uniform thickness.

If funding is provided as planned, experiments could begin on the NIF around the year 2002. While many different kinds of experiments are planned, in this book we are particularly concerned with those most relevant to the use of ICF as an energy source to produce electricity. Of course, the most important result would be the achievement of ignition, to prove the feasibility of ICF.

To help us assess what other kinds of information the NIF might provide, let us pause for a moment to consider how a power reactor based on ICF might be designed. We have now discussed lasers as a possible driver, and the design and fabrication of fuel pellets. It remains to discuss the target chamber (see the upper part of fig. 5), in particular one that can withstand about ten microexplosions per second, as is required for a power reactor.

The most interesting feature that distinguishes ICF reactor designs from tokamak designs is the possibility of protecting reactor structures from excessive damage by neutron bombardment. As far as I know, the ICF method for doing this was first proposed for magnetic fusion, inspired by a 1970 meeting at which Nicholas Christofilos, the visionary inventor of the Astron magnetic fusion concept at Livermore, first understood the threat posed by neutrons, especially in devices providing high power in a small volume, as he hoped his Astron would do. Always alert to any challenge to his brainchild, in short order Christofilos wrote a laboratory report, published posthumously, depicting a method for protecting the metal vacuum chamber by a thick stream of liquid lithium flowing in at the top, down along the walls of the vessel, and out through a hole at the bottom, to be recirculated like a continuous fountain. Research on the Astron ceased soon after Christofilos's death in 1972, but for many years an artist's drawing of his "lithium waterfall" languished behind the radiator in my office while plasma energy confinement consumed our attention. But today, neutron damage to the inner "first wall" of the blanket in a tokamak is a lively research topic. While it does not appear feasible to adopt Christofilos's scheme in a tokamak, it may apply to the newer magnetic fusion concepts that we will discuss in chapter 16. Meanwhile, his idea lives on in ICF.

The first embodiment of Christofilos's proposal for a liquid wall in an ICF reactor was the High Yield Lithium Injected Fusion Energy (HYLIFE) design at Livermore. The liquid lithium, which is exposed directly to the DT neutrons produced in each pellet as it explodes, serves to absorb all debris and most of the neutron energy. Thus, most of the energy is deposited in the lithium stream, which serves as a coolant to carry away the heat. Lithium metal melts at 177 degrees Celsius, so liquid lithium can operate at temperatures suitable to produce steam for power generation. Also, neutron reactions in the lithium breed tritium. Thus, the liquid lithium serves all of the functions of the "blanket" in a magnetic fusion reactor. The actual metal target chamber serves only as the structure holding everything together. Protected by the lithium jets, this structure might last for the life of the power plant. Other considerations concern the size of the vessel—the tradeoff between the greater ability of a larger vessel to withstand repeated microexplosions, and the higher cost associated with the larger size.

Since the HYLIFE reactor design was first proposed in 1979, other similar designs have been considered that retain Christofilos's basic idea of a "disposable" wall while avoiding the danger of fire posed by liquid lithium, which is chemically very reactive with air and water. These ideas include a molten salt called "flibe," which contains lithium but is nonflammable, and streams of solid lithium oxide granules mixed with other granular material that flows through the system like sand. All of these ideas are still at the conceptual stage, but they are promising.

While the NIF will not have a blanket of any kind, the information it provides about neutron energies and debris damage at the target chamber wall will help to benchmark the computer programs used to predict these things in an ICF reactor. It is possible that damage in the NIF will be sufficient to require some kind of protective tiles mounted on the wall, as is already common practice in tokamaks.

Because an ICF microexplosion is of such short duration, there is no concern about vapor from the liquid wall returning to contaminate the pellet during a shot, as there would be for a tokamak plasma that operates more or less continuously in a vacuum. The corresponding problem for ICF is the necessity to clear the chamber sufficiently after a shot to prepare for injecting a new pellet and firing the driver again only a tenth of a second later. Again, the NIF will provide data about conditions after a shot that should be useful in benchmarking the models needed to design suitable target and beam delivery systems, which are now only in a rudimentary, conceptual state.

Ultimately, however, the most important information that the NIF is expected to provide for energy-generating applications of ICF concerns pellets and drivers—in particular, pellets and drivers suitable for an economical and safe power plant. For example, a driver that ignites a pellet but is too wasteful of energy is not suitable: all of the electricity produced by fusion would be consumed to operate the driver. Similarly a pellet, or a pellet mounted in a hohlraum, that produces an exciting energy gain is nonetheless useless for energy production if the pellet structure is too exotic and expensive. And finally, all drivers are not suited to all pellet designs.

As a case in point, let us return to the situation that prevailed when I served on the FPAC committee in 1990. At that time, despite their great success in the laboratory, glass lasers were considered to be out of the race for ICF energy production because of their inefficiency. Two committees of the National Academy of Sciences, in 1986 and again in 1990, had suggested that a totally different kind of driver, not mentioned above, was likely to be the driver of choice for ICF energy applications. This is the "heavy ion" driver, which at that time was receiving modest funding from the Department of Energy purely for the energy application. Our committee recommended that this funding be continued and increased.

The heavy ion approach, actively pursued in the United States at the Lawrence Berkeley National Laboratory, is one of two ion accelerator projects geared toward producing ICF drivers. The other, mentioned previously, is the "light ion" approach being pursued in the United States at Sandia National Laboratories, in Albuquerque. The heavy ion approach is also pursued at Darmstadt, in Germany.

The attractiveness of ions and other charged particle beams as ICF drivers lies in their high efficiency, the mature state of knowledge of particle accelerators, and the ability to calculate the deposition of particle energy in a hohlraum or pellet from first principles. The first accelerator to be explored for ICF was a high-power electron accelerator thought to be very inexpensive by comparison with lasers. This concept, the first idea pursued at Sandia, drew considerable public attention when a Soviet visitor gave talks on the subject (later quoted in *Aviation Week*) disclosing information that was at that time still secret in the United States. Later, when it was concluded that electrons did not deposit energy optimally, Sandia turned its efforts to light ions, currently lithium ions. While this approach is still a potentially inexpensive way to produce high energy, thus far it has not been possible to focus this energy on a tiny target in the manner required for ICF.

The heavy ion approach at the Lawrence Berkeley National Laboratory is based on a different kind of accelerator, originally developed by Christofilos to provide high-current electron beams for his Astron concept. This is an "induction" accelerator in which, by Faraday's law of induction (see chapter 6), a rapidly changing magnetic field induces a voltage that accelerates the ions, which in this case are heavy ions such as cesium, which deposit their energy in a target with good efficiency. An important problem for this approach, and the thesis topic of one of my graduate students, is how to focus such an ion beam onto a tiny ICF target, especially given the environment inside the target chamber in the aftermath of the previous shot. From such considerations it is assumed that a heavy ion driver is best suited for indirect drive. Thus we begin to see how driver and target combinations may become linked as deeper consideration is given to ICF as an energy source.

To ensure uniform irradiation of the pellet, the target must be positioned in the path of the laser or ion beams with great accuracy. For present experiments, and in the NIF, the target is placed mechanically, ahead of time, and a weak "tracer" laser beam is used to locate the target and verify that it is in the proper position. In a power reactor, the targets must be injected several times a second; the main laser or ion beams would be fired only when tracer beam measurements indicated that the target was ready. The tracer information would be used to readjust the aiming of the driver to hit the target, a readjustment that is especially easy with ion beams steered by magnets. The necessary adjustments in aiming are minuscule, and there is plenty of time to gather information, since the target is moving slowly compared to the time required for computers to process information from the tracer laser beam.

There is also the matter of the cost of fabricating pellets. The pellets for present-day ICF experiments (and for the NIF) are made individually and require extensive hand labor. Targets are individually scrutinized during assembly to ensure that they meet standards. This process involves delicate measurements to determine such properties as material composition (by x-ray spectroscopy), surface roughness (by interferometric measurements from several angles and atomic force microscopy), and coating thickness and uniformity (by electron microscopy). As a result, the average cost of a pellet in current ICF experiments is ten thousand dollars, whereas the pellets mass-produced for electric power production must cost no more than a dollar each. System studies indicate that such cost reductions are achievable using automated fabrication techniques similar to those employed in the electronics industry.

Finally, in indirect drive the hohlraum surface, to produce x-rays efficiently, must be a heavy metal, and it may be necessary to recycle the material to permit the removal of radioactive waste created by neutron bombardment during the shot. As in magnetic fusion, wise choices of materials can reduce radioactive waste issues substantially.

Fortunately, whatever pellet and driver combinations turn out to be best for ICF energy production, the NIF appears to be designed with sufficient flexibility to explore the necessary physics and to benchmark the ICF computer models over a sufficiently wide range of parameters to validate them. Nonetheless, beginning to sort through the target designs suitable for cheap mass production and identifying efficient drivers matched to these targets would seem to be an urgent matter if NIF information is to be used expeditiously for the energy goal for ICF. It was just this kind of guidance we had in mind, back in 1990, when our FPAC committee recommended a stronger identity and urgency for the energy component of ICF research, at that time just a fledgling among a variety of energy-related science programs. To make the point, in recommending more funding we suggested a new name, Inertial Fusion Energy (IFE), for this part of the work. We got the name, but not the money. For, as we were soon to learn, while support for the NIF would remain strong for its other purposes, energy was not high on the federal agenda.

IV

FUSION AND POLITICS

14 : The Fusion Dilemma

. . . .

I first heard of "cold fusion" in 1989, in a telephone call from a CNN reporter. It seemed there had been a big breakthrough in fusion, soon to be announced, and would I comment on it? When I asked what the breakthrough was and who had made it, the reporter said that this information was embargoed until the big announcement. Highly suspicious, I declined to comment.

All became clear the next day, when by chance I happened to visit Livermore as a consultant, as I had done periodically since my departure to join the faculty at Berkeley. Apparently there had indeed been some kind of announcement on television claiming the production of fusion energy in a simple chemical cell containing paladium metal, and as a result the Livermore patriarch, Edward Teller, had hastily called a meeting, which I attended.

What occurred in the days after this announcement of "cold fusion" by Stanley Pons and Martin Fleischmann is an interesting example of the scientific attitude. I have heard it said that the difference between a scientist and an astute politician is this: In the political world, the response to a startling claim should be, what did they say? Why did they say it? Why did they say it *now*? But in the scientist's world, as in the case of cold fusion, the main consideration is, can it be true? And so there we were, about twenty of us, huddled around a videotape player rerunning the cold fusion announcement over and over, trying to figure out what it could mean. Teller was so excited that at one point he suggested that I drop everything at Berkeley to return to Livermore and lead an effort to pursue whatever this new thing was.

By noon, about the only information we had gleaned from the video replays was the brand name on a few storage batteries used to provide power that was supposedly amplified in passing through the cold fusion palladium cell. Then, in the lunch line, I ran into old acquaintances who said that they had been in their laboratory all night trying to reproduce the Pons and Fleischmann claims, with no success. So it turned out all over the world.

For my part, as a theorist, I kept thinking about the new claims until finally, unable to put them out of my mind, I went into my office on the following Sunday and began digging through old books like some medieval alchemist, searching for any clue in the quantum mechanical "tunneling" theory that could possibly allow deuterium atoms dissolved in palladium metal to undergo fusion reactions, as Pons and Fleischmann claimed they had demonstrated. Absorbed and perplexed, I almost missed Easter brunch with my family when, now a bona fide absent-minded professor, I managed to lock my car and office keys inside my office as I was leaving.

Within a month I had been drafted at the annual meeting of the National Academy of Sciences to serve on a panel that the Department of Energy had called together to investigate the claims of Pons and Fleischmann. By now there were many doubters who had tried unsuccessfully to reproduce cold fusion, and the experimental evidence offered to our committee by cold fusion's proponents was crude and unconvincing. For me, the clinching point was the absence of the two byproducts, tritium and helium, that would be expected to occur simultaneously from the fusion of pure deuterium. Without some convincing evidence of a nuclear process that could produce vast amounts of energy, the other claims of intense heat produced by the reaction were unlikely to be of much interest, as Pons and Fleischmann themselves had apparently appreciated in postulating a fusion origin for their observations.

Our committee was negative about cold fusion, as is well documented in our report and in a later book by the committee co-chair John Huizenga, of the University of Rochester. Though a few scientists and a coterie of other enthusiasts have continued to claim otherwise, as far as I can tell the overall conclusion that cold fusion is an illusion still prevails, though I confess I have not been motivated to pay much attention lately. I did, however, experience firsthand the excitement that cold fusion has engendered among believers when, as late as April 1994, a young congressional aide, whose eyes had glazed a bit as I tried to explain the tokamak, eagerly pressed into my hand a new cold fusion magazine published by a constituent back home.

Of course, the brief excitement about cold fusion lay in the possibility of a cheaper, quicker path to fusion energy. Nonetheless, it did show that, despite the apparent disinterest in Washington and among the public concerning energy, the old allure of fusion as an abundant and environmentally attractive energy source for the twenty-first century is alive and well. Even President George Bush, just to make sure something wonderful was not being overlooked, invited Nobel laureate Glenn Seaborg to Washington for a personal assessment of cold fusion.

One might have thought, therefore, as some advisers to the fusion program used to assume, that the actual demonstration of fusion energy production at Princeton beginning in 1993 would have opened the door to the U.S. Treasury. Yet, the fact is that the United States has for the most part been reducing its budget for magnetic confinement fusion for more than a decade, even as Europe and Japan have increased theirs, and the purely energy-oriented component of inertial confinement research has remained exceedingly small.

Maintaining enthusiasm for fusion research among government officials has never been easy. The sustaining force has primarily been the fusion scientists themselves, supported by a few visionary leaders in the administrative and legislative branches of various governments who did what they did with little expectation of political rewards. The late Donald Kerst, one of the founding fathers of fusion research, once said to me that he feared that fusion would never succeed because of the difficulty in passing on the task from one generation of scientists to the next. Half joking, I had reminded him of the Cathedral of Chartres, a Gothic masterpiece four centuries in the making, and observed that fusion, too, would be an enduring edifice. Indeed, if one counts as one generation the ten or so years of college and graduate studies required to earn a Ph.D. degree, I was of the second generation and my graduate students are of the fifth; and bright young undergraduates still come to me eager to become a part of the fusion enterprise. Yet, the challenge that attracts these students, and professional scientists as well, the challenge of advancing a new science step by step, has not always made fusion attractive to those who can see only how far we must go, rather than how far we have come.

The first person in the United States to deal seriously with the dilemma posed by the long, costly development process required to achieve fusion was Robert Hirsch, the director of the U.S. magnetic fusion research effort during the energy crisis of the 1970s. Seizing on the public interest in energy, Hirsch had tried to obtain a governmental commitment to the goal not only of establishing the scientific credibility of fusion but also of implementing an actual demonstration of the production of electricity by fusion that might win the support of industry. The precedent was President Kennedy's decision to place a man on the Moon within a fixed period, ten years—a move that focused and advanced space exploration as nothing else could have done. A few years later, I along with others appeared in a privately funded documentary film in which the astronaut Neil Armstrong said that a country that could send him to the Moon could surely develop fusion.

While Hirsch's insistence that the Tokamak Fusion Test Reactor should be a DT-burning experiment proved to be a brilliant choice, two obstacles to the remainder of his strategy immediately beset his successor, Edwin

Kintner. These obstacles have for the last twenty years shaped the history of magnetic fusion development in the United States, and indeed the world. Thus, before trying to assess the future, I shall devote the remainder of this chapter to a review of these issues and the ensuing history.

The main issue that Kintner faced, as soon as the Carter administration arrived on the scene in 1977, was questioning of Hirsch's self-imposed urgency in developing fusion. Even though it was President Carter who established the Department of Energy, Carter and his advisers did not agree with their predecessors' strong emphasis on nuclear energy; they wanted to halt the fission "breeder reactor" and divert some of the fusion funding to projects, such as "clean coal," that might have an earlier payoff. The second issue, a corollary of the first, was a concern that Hirsch's rapid advance of the tokamak would freeze out other ideas that might make a better reactor in the long run.

Kintner's resolution of these issues, encouraged and articulated by a high-level review chaired by the late Solomon Buchsbaum, was the promotion of a dual path of development in which the tokamak had a central role, but that in principle also served the needs of other magnetic confinement schemes to be developed in parallel. The centerpiece of this strategy, more or less adopted later in other countries, was the tokamak not as the final power reactor concept but as an engineering test facility that would advance nuclear engineering for any confinement scheme. As we have seen, Kintner's approach survives today as ITER.

Kintner's strategy followed the classic approach—like the technique used by a cook preparing a four-course meal—of separating the problem into individual components, or program elements, that could be carried out independently and only later coalesced into a practical power reactor. As in cooking, different processes would proceed on different timetables so that everything would be finished on time. As mentioned above (see chapter 10), Kintner was especially intrigued by the complete separation of thermal power production—in the thick blanket that surrounds the burning plasma—from the production of fusion energy in the plasma itself. It was this fact, remember, that justified proceeding with an engineering test facility based on the tokamak as the plasma source while continuing to develop other confinement schemes in parallel.

The main beneficiary of Kintner's strategy during his tenure as director was the mirror program at Livermore. Kintner became involved with the mirror program immediately after becoming Hirsch's deputy in 1975, and he became more involved when he succeeded Hirsch as director of the U.S. magnetic fusion program a year later. This was the period during which the Livermore mirror team was scoring a succession of breakthroughs in their 2XIIB neutral-beam injection experiment (see chapter 9), "pulling rabbits out of the hat" one after the other. Kintner soon became a strong supporter of our proposal, initiated under Hirsch, for the building of the Mirror Fusion Test Facility. Originally the MFTF, like the 2XIIB itself, was to be a yin-yang mirror (see chapter 9). However, Kintner was dissatisfied with the prospects of a yin-yang mirror as a future reactor. I could not disagree with this, as I myself had concluded, in a paper published in 1966, that the theoretical maximum yield of such a device was a fusion power equal to the neutral-beam injection power, or break-even. Even so, Richard Post, at Livermore, had shown that one could still produce net electricity from the fusion power by cleverly recovering the neutral-beam power so efficiently that very little net power was consumed by the neutral beams. But Kintner wanted something better and exacted from me a promise that, as his price for approving the MFTF, we would try to seek improvements. This became my chance to be an inventor.

Many people had, of course, been thinking about how to improve the power gain beyond break-even in a mirror device, especially since the 2XIIB had begun giving encouraging results. Post and I used to talk about it every day while car pooling to my home in nearby Walnut Creek. Then, only a few months after Kintner's challenge, something "clicked" for me. On the evening of July 4, 1976, I scribbled the ideas and numbers that became the "tandem mirror" just before leaving to view the holiday fireworks.

The original idea, simple in concept, yielded on paper about a fivefold improvement over a yin-yang magnetic mirror, roughly what Post had hoped for a mirror device in the early days before calculations such as mine had shown break-even to be the best possible result. The problem was a buildup of positive voltage that occurred naturally in order to confine electrons in a yin-yang, since in the absence of this voltage collisions would cause the electrons to leak faster than the ions. It was the fact that positive voltage also ejected low-energy ions that helped to degrade mirror confinement to the break-even level. But in the tandem mirror, the positive voltage was put to good use.

For me, the tandem mirror idea followed directly from contemplating the 2XIIB yin-yang mirror results, and also from international cooperation. At first the 2XIIB experiment seemed to have failed: the ion confinement time was only getting worse due to instabilities as the machine was made cleaner by improvements in the vacuum system. Then, in the nick of time, a polaroid picture arrived with a Soviet visitor, sent to us by Ioffe, of "Ioffe bar" fame, who we knew was pursuing Post's idea that a stream of cold plasma flowing through the yin-yang mirror would control instabilities. This little picture showed that Post was right.

Soon thereafter, the Livermore team leader, Frederick Coensgen, developed a "plasma stream" for the 2XIIB, and by June 1975 the 2XIIB was making history. But it had been difficult to force the stream to flow through the 2XIIB yin-yang mirror, because of the positive voltage that repelled the ions in the stream. It was this detail that I was thinking of in July 1976 when the light dawned. If ions would not pass through the positive voltage of the yin-yang, then the yin-yang could be used like a dam. This was the idea behind the tandem mirror, which consists of three components—a long solenoid with yin-yangs at each end. Injecting neutral beams to maintain a positive voltage in the yin-yangs dams up ions in the solenoid, even at fusion temperatures if the voltage is high enough. Most of the fusion power would be produced in the long solenoid, the yin-yangs serving only to dam up the ends in order to maintain good plasma confinement in the solenoid. Limited to break-even, the yin-yangs actually consume power, but much less than that produced by a solenoid of sufficient length.

While I was on vacation, another member of our group, Grant Logan, came into Post's office with the same idea. Post showed him my notes. By the time I returned from vacation, there was considerable enthusiasm around Livermore, even to the extent of dismantling our Baseball II facility to make room for a new Tandem Mirror Experiment (TMX). Then, when we learned at an international meeting in 1976 that a Soviet, G. I. Dimov, had also thought of the tandem mirror, Kintner was so convinced that he virtually approved the TMX in public even before we had proposed it officially.

The TMX was completed by 1979. A Japanese version had been completed at the University of Tsukuba before that, and Dimov began building a tandem mirror at Novosibirsk. Soon Logan and David Baldwin improved on the idea, yielding a theoretical energy gain of twenty times break-even. Finally, armed with positive results from the original TMX and encouraged by Kintner, researchers at Livermore went on to propose, in 1981, that the MFTF, originally a single yin-yang, be extended straightaway to a tandem mirror configuration, the MFTF-B. In 1980, Logan won the prestigious Ernest O. Lawrence Award for his part in the tandem mirror invention.

Yet all was soon to come to naught. Efforts to implement Kintner's strategy had run their course by 1986, when the mirror program at Livermore was canceled, the Mirror Fusion Test Facility (MFTF-B) was officially "mothballed," and Livermore accepted a modest role in the tokamak program. Kintner himself, later to be a prominent executive in the utility industry, had resigned in protest in 1981 when he felt that the Department of Energy had failed to back him up in a dispute over MFTF-B with the Office of Management and Budget. Soon after Kintner's resignation, the U.S. magnetic fusion budget, after reaching a peak of $471 million in 1984, declined about $50 million per year for three years, leading to the loss of the three major projects Kintner had initiated: the Fusion Materials Irradiation Test facility; the Elmo Bumpy Torus (EBT) concept improvement project at Oak Ridge; and finally the MFTF-B. Of these, only the MFTF-B had been fully constructed. Nearly the length of a football field, and crammed with superconducting magnets, cryogenic-vacuum systems, and other avant garde technology, the MFTF-B was a brilliant engineering accomplishment by project manager Victor Karpenko and his team, many of whom went on to careers in other large, science-related construction projects. When the facility was officially canceled, the very day of its dedication ceremony, we had a party anyhow, to honor the engineers who built it.

While scientific issues played some part in specific decisions, and many unanswered questions remained concerning mirror physics in particular, the primary motivation for these changes in the program was financial, as was made clear by Alvin Trivelpiece, who was still director of research in the Department of Energy at the time the budget cuts were implemented. Despite these wrenching changes and its failure to win approval for any new construction project since 1980, the U.S. magnetic fusion community has largely continued to espouse Kintner's "dual-path" development strategy, using the tokamak in its present form to advance toward ignition and nuclear engineering, while continuing to improve the tokamak and develop other ideas in smaller experiments in parallel. The Fusion Policy Advisory Committee on which I served in 1990 endorsed this view, which was then adopted in 1992 in President Bush's National Energy Plan, which set a target date of 2025 for a fusion Demonstration Power Reactor. Key recommendations included U.S. participation in the ITER Engineering Design Activity, the successor to Kintner's Engineering Test Facility; a materials development program "moving toward" an international test facility like Kintner's FMIT; and a concept improvement program, though now mainly focused on improving the tokamak itself by making it steady-state by means of non-inductive current drive (see chapter 10).

When I was writing this, in June 1995, a new fusion advisory panel under the auspices of President Clinton's Committee of Advisers on Science and Technology (PCAST) had just issued a report that again acknowledges the need for all of the above program elements if the fusion program is to meet the National Energy Strategy goal of an electric power demonstration by 2025; but the PCAST report expresses doubt that the necessary funding—about double the 1995 level—would be forthcoming. Internationally, two large new projects have been initiated, one in Germany and one in Japan, both being stellarators intended as concept improvement projects.

In the mid-1980s, Kintner's successor, John Clarke, led an effort to rejuvenate the U.S. magnetic fusion program around the nearer-term objective of ignition, which he saw as the natural follow-on to proof of scientific feasibility or achievement of break-even in the TFTR and JET. Of course, ignition, or near-ignition, had been a subsidiary goal of Kintner's Engineering Test Facility and would become the main goal of the first phase of ITER operation, the physics phase. What Clarke and the other U.S. fusion leaders had wanted was the least expensive way to achieve ignition as an objective in its own right.

It was to the Alcators, a series of small tokamaks at MIT, that Clarke and others turned. Thinking back to the tokamak design rules developed in chapter 2, we would find that the key to a small device is a very strong magnetic field. In the early 1970s, Bruno Coppi and others at MIT had built the first Alcator tokamak to exploit this fact using compact magnet design techniques developed by Francis Bitter. Later Robert Bussard founded a short-lived company to produce "throwaway" tokamaks based on this idea. However, Coppi's main motivation had been to devise a relatively inexpensive way to study the physics of tokamaks in a university environment. By the early 1980s, having succeeded brilliantly with a series of Alcators, Coppi had begun to champion a larger Alcator as a cheap way to reach ignition. Coppi's idea inspired Princeton, with strong backing from Clarke, to propose the Compact Ignition Tokamak (CIT). The CIT was intended to burn by itself for just the few seconds required for the heat emitted by the fusion alpha particles to come into balance with the heat leakage from the DT plasma.

The CIT soon captured the enthusiasm of American fusion scientists. At the 1986 ceremony at which he was awarded the very prestigious Fermi Prize, senior theorist Marshall Rosenbluth offered to forego the considerable cash prize in return for a pledge from the secretary of energy, John Herrington, to "make CIT happen." For the moment, the dual path took second place to a total concentration on advancing toward ignition, even though few if any regarded the CIT design as a direct path to a reactor. My own evidence for this attitude came in a meeting of my mirror advisory committee, a group of industry representatives who had long supported the tandem mirror as a more desirable reactor configuration than the tokamak. Yet, when I asked them whether ignition in CIT should have higher priority than the less-developed mirror, all chose ignition as their first priority. Meanwhile, by 1990 the inertial fusion community had begun to formulate its own ignition experiment, which became the National Ignition Facility.

By mid-1995 the NIF, with its different financial sponsorship under Defense Programs in the Department of Energy and its near-term mission for that sponsor, appeared to be moving forward, while the CIT had been abandoned in favor of the less expensive Tokamak Physics Experiment (TPX), which was aimed at achieving the steady-state goal called for by the FPAC back in 1990. To some extent, the CIT had turned out to compete with the emerging ITER project, though we Americans always saw it as a nearer-term, domestic project that was supportive to ITER but not a substitute for it. The 1990 FPAC report had recommended that the United States "take an even-handed approach in strengthening its national and international efforts" by not only participating in the ITER Engineering Design Activity but also constructing a slightly scaled-down version of CIT, called the Burning Plasma Experiment (BPX). The 1995 PCAST report made it clear that the real issue was whether or not any magnetic fusion construction project—be it ITER, TPX, or anything else—could be initiated by the United States in the foreseeable future. Indeed, as we shall see, by 1996 the TPX, too, was gone.

Thus, once again, we come full circle to the central political dilemma of fusion research—namely, how to sustain governmental support for a long and challenging technical endeavor. Sizing up the situation in 1981, at the beginning of what was to be a long dry period of declining budgets and unsuccessful bids for new facilities, Joan Bromberg, in the last chapter of her history of the magnetic fusion program, speculated about various reasons why fusion research might or might not gain the support it would need to carry through. Some of these have now been tested. Above all, her warning that "success in fusion . . . has never been viewed as assured" can now be considerably modified, especially in regard to "the crucial and thus far unattained regime of burning plasmas." As I have tried to stress in parts I and II of this book, our theoretical understanding and experimental exploration of plasma phenomena, including DT experiments, the "regime of burning plasmas" to which Bromberg refers, have made enormous strides. Perhaps more than is widely appreciated, the DT-burning experiments in TFTR have pierced the veil of ignition physics even beyond our expectations. As we have seen, the two crucial elements of ignition, as the term is used in magnetic fusion, are the confinement of the energetic alpha fusion products long enough to heat the DT fuel; and an energy confinement time and pressure for the fuel large enough so that the alphas alone can heat the fuel plasma to the temperatures required for fusion reactions to occur. In its original design, TFTR took no credit at all for alpha confinement and heating. Yet, as we saw earlier, the energetic alpha particles in TFTR are actually confined as well as they need to be in a much larger ignition experiment such as ITER. The reason that the smaller TFTR has not already ignited concerns the energy confinement in the fuel, not the alphas. Thus, much of what we need to know to design an ignition experiment with high confidence is already being studied in TFTR.

Yet, the fact remains that despite these very encouraging results in TFTR, the budget has not responded, and the fusion dilemma continues. The PCAST panel stoically advised researchers to go on doing the best they could with TFTR, the DIII-D, and the other devices in hand while trying to negotiate once again a stretched-out program, heavily international, that governments might be willing and able to fund. As the PCAST report notes, this appearance of vacillation on the part of the United States is "straining the patience of this country's collaborators in the international component of the fusion effort"—though in fact not only the PCAST report but also Congress and the Department of Energy have thus far stressed the commitment of the United States to the completion of the ITER Engineering Design Activity.

This bleak outlook might change if the public's interest in energy was reawakened or if a technical breakthrough occurred that could significantly reduce the timetable for fusion development. In the last analysis, the magnetic fusion budget initially rose to the level it did as a direct response to the unanticipated oil crisis of 1973 and the program's technical readiness to leap forward following the Soviet breakthrough with the T-3 tokamak. Though the U.S. fusion budget in as-spent dollars peaked in 1984, in constant dollars the peak year was actually 1977, the final year of the rapid fivefold increase that followed the 1973 oil crisis.

What might trigger a new crisis, and would such a crisis matter? Almost surely, any new urgency to accelerate fusion research and research on other energy alternatives would result from a threat to the environment that finally convinced U.S. and world leaders that a deliberate shift away from fossil fuels was essential for the well-being of society. This kind of world response began to surface in the late 1980s, with heightened attention to "global warming." While it would only be speculation for me to guess when and whether these or other environmental concerns will recur and intensify, it is appropriate for me to mention the environmental merits of fusion, which I detail in chapter 15.

As to fusion breakthroughs, I am happy to report that, budget gloom notwithstanding, the irrepressible curiosity of fusion scientists continues to probe in new directions, encouraged by a new confidence based on all that has been learned with the resources generated by the earlier surge in fusion funding. In chapter 16, we will look at three areas for potential breakthroughs, two magnetic and one inertial, that could make a major difference.

Finally, I should point out that on the whole, fusion research has managed to maintain its funding better than any of the other energy research projects undertaken in the 1970s. Fusion has been supported both as a potential new energy source for the next century and also as a pioneering science with the kind of spin-offs one hopes will arise from government-sponsored science. Companies involved in fusion, and new companies and consortia spun off from fusion laboratories, are leaders in producing the equipment used in making computer chips; in hardening wear-resistant industrial tools and parts; in creating protective coatings to resist corrosion; in lighting; in the manufacture of biodegradable wrapping and packaging; and in devising plasma devices to dry inks and coatings and to selectively destroy toxic chemicals in waste treatment plants. Companies involved in building magnetic fusion devices have used their expertise to market the magnetic resonance imaging (MRI) systems now used as diagnostic tools in most hospitals; magnetic energy storage systems; magnetic separators used in industrial purification of Kaolin clay; and industrial equipment for use in the "magneform" process that shapes metal parts ranging from baseball bats to airplane wings. This list of products and processes could go on and on, and fusion research has also led to advances in basic plasma theory and computation applicable to space physics and astrophysics.

The good science that is integral to fusion research, and fusion's ultimate energy potential, are the factors to which the scientists themselves have most often appealed when seeking support, and which have probably preserved fusion through the many changes in economic, political, and social conditions that have occurred during its long history. Like any other science funding, this traditional kind of support for fusion is driven less by timetables for the construction of a demonstration plant than by the need to maintain scientific progress.

The recent PCAST panel report puts this traditional case for supporting fusion very well in its opening paragraph:

> Funding for fusion energy R&D by the Federal government is an important investment in the development of an attractive and possibly essential new energy source for this country and the world in the middle of the next century and beyond. This funding also sustains an important field of scientific research—plasma science—in which the United States is the world leader and which has generated a panoply of insights and techniques widely applicable in other fields of science and industry. And U.S. funding has been crucial to a productive, equitable, and durable international collaboration in fusion science and technology that represents the most important instance of international scientific cooperation in history as well as the best hope for timely commercialization of fusion energy at affordable cost. The private sector cannot and will not bear much of the funding burden for fusion at this time because the development costs are too high and the potential economic returns too distant. But funding fusion is a bargain for society as a whole.

Often, society needs leadership to make it aware of the bargains available to it, especially for the long haul. Far better than a rekindling of energy awareness at a time of crisis would be foresight on the part of leaders who saw the need for new and better energy sources in time to avoid a crisis. I opened and closed part I of this book with the familiar story of Prometheus, who saved humankind by bringing the fire of the Sun to Earth to allay the unintended crisis created by the poor planning of his well-meaning brother, Epimetheus, whose name means "afterthought." The magnitude of the trillion-dollar electricity generation and distribution system in the United States alone, and the critical role it plays in our industry, our national security, and our private lives, are such that no Prometheus can be expected to solve a true environmental energy crisis overnight. However, most likely the crisis will not be instantaneous; rather, it will come upon us gradually, perhaps going unnoticed until it is too late to avoid serious consequences. Without the vision to prepare itself with better alternatives, society will simply continue to pollute in whatever way is necessary to provide its needs for electricity as it sees them. For example, the People's Republic of China, which in 1990 included one-fifth of the world's population but was responsible for only one-eleventh of the world's annual energy consumption, is now experiencing and planning a rapid increase in its economy tied to increasing energy and electricity production. According to Lu Yingzhong, director of centralized energy planning, electricity consumption in China increased a hundredfold from 1949 to 1987, and even with extensive conservation measures, by 2050 per capita energy consumption in China could reach or surpass the already high levels that developed countries had ten years ago. Coal, the energy resource most abundant in China, will provide most of this growth.

The need for forward-looking leadership in the energy arena was the main conclusion reached by an international seminar on electricity supply and demand that I co-hosted in Berkeley in January 1990, a year before the oil-inspired Persian Gulf War. The meeting was attended by thirty-two experts in energy finance, management, and technology from twelve countries around the world, in North and South America, Europe, and Asia. The seminar group conveyed its findings and conclusions to various governmental bodies, stressing the vital role of electricity to societal well-being and the need for global action "to turn public attention to the emerging problems of electricity shortfalls in many countries and their dramatic implications for health, employment, economic hopes, and the environment."

But, as Kenneth Davis, former U.S. deputy secretary of energy, warned at the news conference following the seminar, "our recommendations will come to nothing unless we can get the attention of leadership, and I mean both nationally and internationally. And by leadership, I mean that it is vital that the President of the United States take a strong position on the need for adequate supplies of energy to sustain our economy. Similarly, leadership is needed from other countries. In this country this also means courage in the Congress and also from the leaders of the electric utility companies who are responsible for supplying the electricity."

That was more than five years ago. And the clock keeps ticking.

15 : Fusion and the Environment

· ·

To speak of fusion and the environment is to invite a comparison between fusion and fission. Since there are many fission reactors currently in operation and coal is still abundant in the United States and many other places, it is environmental issues, rather than inadequacy of fuel supplies, that will most likely bring fusion into the marketplace. Both fission and fusion are likely to be more friendly to the environment than their most probable competitor for electricity generation, namely coal—a point quickly grasped by the noted American conservationist and photographic artist Ansel Adams, who strongly supported fusion before his death in 1984. One of his last photographs, included in his posthumous autobiography, is a picture of the laser fusion target chamber at Livermore made to show his support for fusion research. He took this photograph, and also shots of our yin-yang magnet, when visiting the laboratory at my invitation, after he called me one day to express his frustration at an antifusion article and to ask what I was going to do about it. "Well, what are you going to do?" I replied. And so he came to do what he did best.

Fusion research first attracted the support of U.S. environmentalists during the 1970s, when the Atomic Energy Commission was heavily committed to fission—in particular, the breeder reactor—as the long-term answer to energy needs. To some degree, fusion was being used to fight the breeder reactor, which had become the focus of antifission sentiment. Newspaper articles began to appear describing fusion as "virtually pollution-free" and calling for large increases in funding for fusion research.

By the mid-1970s, when fusion funding was already on the rise, a few far-thinking scientists began to attempt serious comparisons between fusion and fission, since by this time they realized that whatever environmental advantages fusion might have were indeed the primary rationale for the support of fusion as a practical power source. One of these scientists was John Holdren, who spent many years on the faculty at Berkeley.

I first met Holdren when I hired him at Livermore as a young plasma theorist fresh out of Stanford. In addition to doing plasma research, he had also been a protégé of Paul Ehrlich, author of *The Population Bomb* and other popular books on matters of environmental concern; Holdren was co-author of two of these books. Because of his broader interests, Holdren soon moved on to environmental work at the California Institute of Technology, but we quickly renewed our relationship when the University of California, Berkeley, hired him in 1970 as the first appointee to its new, nondepartmental Energy and Resources Group, and I engaged him as a consultant at Livermore. He currently serves on the President's Committee of Advisers on Science and Technology and chaired the PCAST fusion panel that produced the 1995 report quoted in chapter 14.

In 1985, the Department of Energy's Office of Fusion Energy asked Holdren to chair a national committee of fusion and fission experts to take a fresh look at the environmental, safety, and economic aspects of magnetic confinement fusion reactors. This study, known by the acronym ESECOM (pronounced "Easy-Com"), has served as a standard reference ever since its publication in executive summary form in 1988 and the publication of the complete report in 1989. After I moved to Berkeley in 1988, Holdren and I, with several students and other faculty members, continued to work on various topics following up the ESECOM study. Much of what follows reflects what I have learned from these studies with Holdren and our students.

Another aspect of my Berkeley experience that influences what I write in this chapter is my seven years as chair of the Department of Nuclear Engineering, long known for its research on the disposal of nuclear waste from fission reactors and on safety issues related to these reactors. As department chair, I helped to organize a new Center for Nuclear and Toxic Waste Management that involved not only engineers but also professors and students from the Energy and Resources Group, the Political Science Department, and the Graduate School of Public Policy. I am deeply indebted to these colleagues, also, for whatever I have learned about issues of public trust and confidence in technology in modern society.

To set the stage for what follows, I should pause to remind the reader of the differences between nuclear fission and nuclear fusion, and the way in which these differences influence the environmental and safety features of fission and fusion reactors.

First consider fission. The nuclear fission of uranium or plutonium invariably produces lighter radioactive elements known as "fission products," and also some radioactive actinides even heavier than uranium. It is these fission products and actinides, some of which remain radioactive for thousands of years, that must eventually be disposed of as nuclear waste. Also, the heat continuously produced by the radioactive decay of fission products could potentially cause an accident by overheating unless the reactor's cooling systems continue to operate properly even after the reactor is turned off. It is these properties of fission products and actinides that have led to political debates over waste disposal sites, and the desirability of a large containment building to prevent the escape of fission products in the event of an accident. Finally, in certain circumstances rapid overheating can occur during fission reactor operation if control systems fail to prevent the fission chain reaction from exceeding the normal "criticality" condition.

Turning to fusion, we find what appears to be a qualitatively different situation. "A thermonuclear reactor is probably an extremely safe device," wrote Albert Simon in his 1959 book on fusion, "since the total fuel in it at a given time is only enough to sustain the reaction for a few seconds and since there are no fission products such as occur in uranium reactors." It was thought that the only volatile radioactive substance of consequence in the fusion process would be tritium, which is among the least toxic of radioactive substances; the artificial radioactivity produced by neutron bombardment of metals in the blanket would be safely locked up in these solid metals.

By the 1970s, scientists such as Holdren had begun to adopt a more cautious tone, as a result of the first studies of the actual consequences of tritium production in the blanket of a DT fusion reactor and the creation of artificial radioactivity by fusion neutrons bombarding the steel or other structure holding the blanket together. "Fusion, like all other energy sources, will not be completely free of environmental liabilities," Holdren and I wrote in a paper we published with Richard Post in 1975, "but the most obvious of these—tritium leakage and activation of structural materials by neutron bombardment—are susceptible to significant reduction by ingenuity in choice of materials and design." In other words, the potential environmental advantage of fusion will not occur automatically: one must work for it. In a thorough review of fusion reactor design published in 1981, Robert Conn was already listing "favored" and "less favored" construction materials, in part basing his classification on low radioactivity.

In 1985, under Holdren's chairmanship, the ESECOM group of fusion and fission experts set out to update the comparison of fusion and fission to see how fusion designs were actually faring. They decided to examine ten different fusion reactor design studies and four fission reactor designs, including one type of fission reactor now operating and three that might be representative of fission reactors competing with fusion in the middle of the next century.

The starting point was a detailed comparison of the amount and types of radioactivity present, depending on the reactor design and the materials used to construct the reactor's various components. This information was available for the fission reactors, but generating comparable data for the fusion reactor designs required hours of computing time on supercomputers in order to do a statistical count of nuclear reactions produced by neutron encounters with the original blanket material, then keep track of any other neutrons produced and their interactions, then calculate neutron reactions with the products of the first reactions, then follow in time the decay of any products that were radioactive, and so on, finally storing all this information so that one could calculate at any chosen time the amount and the nature of radioactive emission of alpha particles, beta particles, and gamma rays. For example, one could calculate the "afterheat" generated after a fusion reactor is turned off, or the decrease of radioactivity with time, to see if there were long-lasting effects as there are with fission products. Fortunately, the computing techniques for doing all this had already been developed to meet the needs of fission reactor designers, and computer programs specific to fusion were already available.

These extensive "neutronics" calculations were required in order to see in detail the radioactive buildup due to neutron bombardment of the blanket. Additional calculations were needed to determine the inventory of tritium in the blanket and other components. Given these inventories of radioactive material, it was then necessary to estimate how much of it would actually escape in the event of a severe accident, what must eventually be disposed of as nuclear waste, and the health hazard in each case. In order to make sense of this wealth of information, the ESECOM group devised various numerical criteria representative of safety and waste disposal characteristics for each system, in each case adapting criteria similar to those developed for fission reactors by the U.S. Nuclear Regulatory Commission (NRC).

Massive exposure to radiation may cause death quickly, typically within sixty days; such deaths are referred to as "early fatalities." Lower levels of exposure, either all at once or over a period of time, may produce longer-term health problems. To evaluate safety, the committee considered both circumstances, one related to the potential for early fatalities at the time of an accident and the other to chronic exposure over a fifty-year period due to radioactive contamination of the land and ground water. In technical terms, the threshold dose of radiation at which an early fatality becomes possible is two hundred rem, the rem being the standard measure of radiation quantity that is used to evaluate industrial hazards, nuclear medical doses, and dental x-rays, as well as exposure due to naturally occurring radiation that reaches us from space or from various radioactive chemicals such as the radioactive potassium present in bananas and all humans. The two-hundred-rem threshold for an early fatality is believed to be well known from actual case histories, while the long-term effects of radiation exposure are more controversial. For instance, the relation between low levels of radiation exposure and various forms of cancer is hard to determine because other causes of cancer are often more common.

One key question for the ESECOM group was whether or not any accident was likely to expose anyone to the two-hundred-rem early fatality threshold dose. For this purpose, as in all of its work, the group took the conservative position of asking what would happen if all of the radioactive material that could be released within the reactor during an accident actually escaped, as would happen if the containment building failed. Moreover, in asking how any released radioactivity would be distributed by the wind, they assumed the worst kind of weather condition—an inversion layer that traps smoke and smog near the ground. Then, as an extreme case, they did mathematical projections of the radiation dose that would reach a person standing, unprotected, for thirty days at a nearby location, taken to be one kilometer from the reactor building. Similarly conservative assumptions were applied to the case of chronic exposure, the criterion being whether a person continuously present at a point ten kilometers (about six miles) from the reactor, never leaving the area, would receive a cumulative dose of twenty-five rem over fifty years (one-half rem per year).

To estimate how much of the radioactive inventory would actually be released, the ESECOM group divided the various radioactive materials into five categories: volatile materials, which would be most easily released (referred to as categories I and II); materials likely to be released if extensive melting occurred during a severe accident (category III); and less volatile materials, which would be less likely to be released even in a severe accident (categories IV and V). Then a "safety margin" was calculated, based on estimates of the proportion of radioactive materials that would be released.

The committee also attempted to assign to each type of reactor a Level of Safety Assurance, by exercising the best judgment of the group based on all of its detailed studies. Level 1, the safest, is intended to mean that no early fatality could occur under any circumstance, either because the total radioactive inventory was too low to create a fatal dose at the site boundary one kilometer away, or because it could be shown convincingly that no accident could release enough of the inventory to cause such a dose. In other words, without any additional emergency response systems, containment building, or operator intervention, the reactor would be incapable of killing any member of the public who happened to be right at the site boundary at the time of an accident. Level 2, almost as good, refers to systems dependent on emergency cooling methods that work automatically and "passively"—for example, natural conduction of heat away from a trouble spot to pools of liquid or air ducts that can take away the heat. Levels 3 and 4 imply progressively heavier reliance on "active" emergency systems that must operate in response to an accident; or, as a last resort, reliance on the containment building.

Finally, to characterize nuclear waste disposal, the ESECOM group first calculated the volume of contaminated material that must be disposed of over the nominal thirty-year life of a reactor. (In the case of fusion, little waste disposal is needed until the plant is finally shut down.) Then a quantity called the "intruder dose" was calculated to determine the exposure to people living and growing their own food at the waste disposal site a hundred years later. To create a uniform basis for comparison, in these calculations the dose is determined as if the nuclear waste qualified for "shallow burial"—that is, as if it could be buried only a few meters below the surface, a disposal technique currently allowed in the United States for "Class C" waste, such as some of the waste generated by hospitals, but not for fission products, which require many more precautions.

Some of the findings published in the ESECOM report are summarized in the table. The fission case shown is the liquid metal breeder reactor mentioned at the beginning of this chapter, which the ESECOM group assumed would be the main competitor to fusion by the middle of the next century. The fusion reactor cases shown include four DT tokamak reactors with different structural material and coolants, and one "advanced-fuel" (deuterium and helium-3) tokamak reactor discussed below. A safety margin of 1 or less indicates the need for active emergency systems, or a containment building not breached by the accident, to avoid exceeding the early-fatality threshold of two hundred rem or the chronic dose of twenty-five rem over a fifty-year exposure. The larger the safety margin, the safer the system. Similarly, an intruder hazard margin less than 1 indicates that "shallow burial" is not allowed. (My "intruder hazard margin" is the reciprocal of the "intruder hazard" calculated by the ESECOM group.)

From the table, we see that all of the fusion cases would have radioactive inventories one thirtieth to one three-hundredth that of the fission breeder in the most easily released categories (I and II), and estimated safety margins against early fatalities or prolonged land contamination due to accidents that exceed similar safety margins for the fission breeder by large factors in most cases. All of the fusion cases shown have safety margins greater than 1, meaning either that no physically plausible release of radioactive materials from the reactor could produce an early fatality or exceed safe limits for ground water contamination, or that the reactor has totally passive safety features that prevent high-temperature excursions that could release radioactive materials otherwise safely bound up in the structure (ESECOM category III materials), giving a high level of assurance that no release could have those results. Since none of these claims relies upon the additional protection that would be afforded by the containment building, the ESECOM group suggests in its published paper that these conclusions show a "high demonstrability of adequate public protection from reactor accidents (no early fatalities offsite), based entirely or largely on low radioactivity inventories and passive barriers to release rather than on active safety systems and the performance of containment buildings." Referring to nuclear waste, the paper goes on to note that, compared to fission reactors, fusion reactors offer a "substantial amelioration of the radioactive waste problem by eliminating or greatly reducing the high-level waste category that requires deep geologic burial."

While the ESECOM studies considered only magnetic confinement fusion designs, the limited amount of work done thus far on conceptual designs for inertial confinement fusion reactors suggests that their environmental and safety characteristics would be similar to those for magnetic fusion. There are differences that require more study, some of which are potentially advantageous, such as the protection of the target chamber structure from direct neutron bombardment by thick liquid or granular layers of material near the wall. Other differences related to the pulsed nature of the energy production (due to the ignition of one pellet at a time) and the materials and design choices in the pellet fabrication system will require attention to detail to minimize environmental impact and accident hazards. On balance, inertial fusion reactors might be expected to compare favorably with the better magnetic designs listed in the table.

Partial Summary of ESECOM Comparison of Fusion and Fission Reactors

| | Tokamaks | | | | | Fission |
	CASE 1 [a]	CASE 2 [b]	CASE 3 [c]	CASE 4 [d]	CASE 5 [e]	Breeder Reactor
Safety						
Radioactive inventories (millions of curies)						
Categories I, II [f]	28	35	33	16	3	1,016
Categories III, IV, V	2,945	2,686	2,243	640	40	4,080
Safety margin	20	60	10	1,000	10	.0007 – 60 [g]
ESECOM Nominal Level of Safety						
Assurance	3	2	1	2	2	4
Nuclear waste						
Total volume over lifetime of the plant (cubic meters)	1,590	2,415	4,030	1,026	730	200 [h]
Intruder Hazard Margin	0.08	0.3	1.8	0.6	24	7×10^{-6}

Source: Holdren, J. P., chair, D. H. Berwald, R. J. Budnitz, J. G. Crocker, J. G. Delene, R. D. Endicott, M. S. Kazimi, R. A. Krakowski, B. G. Logan, and K. R. Schultz, "Report of the Senior Committee on Environmental, Safety, and Economic Aspects of Magnetic Fusion Energy," UCRL-53766, Lawrence Livermore National Laboratory report, September 25, 1989.

[a] DT fuel, vanadium alloy structure, liquid lithium coolant.

[b] DT fuel, ferritic steel structure, helium gas coolant.

[c] DT fuel, silicon carbide structure, helium gas coolant.

[d] DT fuel, vanadium alloy structure, molten salt coolant (pool type).

[e] DHe³ fuel, vanadium alloy structure, water coolant.

[f] DT fusion cases include 10 million curies of tritium.

[g] A wide range for different reactor components, the lowest safety margins representing those containing volatile category I/II materials.

[h] Excluding nuclear waste that would be created in reprocessing fuel elements.

Thus, qualitatively, the situation remains more or less as it appeared in the beginning. If fusion reactor designers do their work with the environment in mind, the very different nature of the nuclear fusion reaction does indeed lead to a new electric power generation system no less safe than the power stations we are used to, with essentially no pollution of the atmosphere. The primary waste disposal issues arise only when the plant is finally shut down after many years of useful life, and fuel deliveries to this plant would be a truckload of nonradioactive deuterium and lithium once in a while—a far cry from the trainloads of coal arriving every day at a large coal-fired power station, with all that this implies in terms of mining, carbon dioxide emission, acid rain, and a massive accumulation of waste.

It remains to be seen whether the attractive features of fusion reactors, even when substantiated by in-depth analysis such as that carried out by the ESECOM group, will fully satisfy the public when the time comes to deploy them. Nor should it be concluded that the greater flexibility in managing safety and waste disposal in fusion reactors implies that fission reactors should not be used when needed. In the 1970s, the chair of the U.S. Atomic Energy Commission, Dixie Lee Ray, advocated the continuing evolution of fission and fusion reactors, believing that fission would continue to be used and improved but that fusion, if successful, would eventually displace fission. Academician Velikhov in Russia has espoused similar views.

In such a scenario, one possible early application of fusion would be to use the neutrons generated by the DT reaction as a competitor to the fission breeder reactor. The purpose of breeding is to extend the use of fission reactors beyond the capacity of naturally available sources of the rare fissile isotopes such as uranium-235, needed for fuel in a fission reactor. In the breeder, neutrons convert certain abundant isotopes into these fissile isotopes—for example, plutonium can be made from uranium-238, or uranium-233 can be made from thorium. The neutrons in fusion reactors could also be used this way, and a fusion reactor using DT fuel could be designed to produce many more neutrons than a fission reactor of comparable power.

This led to the "hybrid" fusion reactor concept, in which a very different kind of blanket, containing thorium or uranium, produces fissile fuel for as many as a dozen fission reactors of the existing type. Since fusion supplies the neutrons, such hybrid reactors would be designed only to produce fissile fuel while avoiding the criticality condition required for much fission to occur. It has been suggested that the hybrid fusion reactor would be a stepping stone toward the reactors we have discussed up to now: "pure" fusion reactors that would provide heat with which to make electricity directly. The hybrid application would provide an early incentive to deploy fusion systems, and practical experience with a few hybrid systems would give researchers confidence to proceed with pure fusion electric power more expeditiously. In 1988, on the only occasion I had to meet the prominent dissident Russian physicist Andrei Sakharov, I asked, "What about fusion?" Without blinking he said simply, "The hybrid," and continued shaking hands with his guests.

Going further, even if fusion reactors using DT fuel are the first fusion reactors to be developed, there is reason to hope that fusion itself will continue to evolve in ways ever more friendly to the environment. Other fusion processes, using what are often referred to as "advanced fuels" in recognition that their use may come later, produce fewer or less energetic neutrons and do not require the breeding of tritium. Their development would move fusion in the direction of less radioactivity while preserving the benefits of abundant fusion energy.

One such advanced fuel is a mixture of deuterium and helium-3 (DHe^3) that is the basis for one of the fusion reactor cases studied by the ESECOM group (case 5 in the table). The deuterium itself does produce some lower-energy neutrons, and helium-3, though not radioactive, is not abundant on Earth, though it could be produced from tritium. Interest in the DHe^3 reaction increased when Gerald Kulcinski, at the University of Wisconsin, drew attention to the fact that NASA astronauts had discovered that helium-3 produced by the Sun is abundantly absorbed in the soil of the Moon. Even before it is economical to bring fuel from the Moon, this could be an ideal source of fuel for interplanetary expeditions using fusion rockets built and based on the Moon. Or so I contend along with Edward Teller and other coauthors in a paper published in 1992.

Another fusion reaction getting renewed attention is the reaction of ordinary hydrogen with boron (pB^{11}). Like the DHe^3 reaction, this does not produce a neutron. However, while its advocates use slogans such as "aneutronic" to imply that it produces little or no radioactivity, I hasten to point out that any nuclear reaction must be assumed to produce some small amount of artificial radioactivity in surrounding structures until it is proven otherwise. Also, the reaction with another isotope, boron-10, produces both neutrons and tritium. Ordinary hydrogen is of course superabundant in water, and boron is even more abundant than the lithium required to breed tritium for DT reactors. There is, by the way, enough lithium in desert brines and the sea to sustain a DT fusion economy for millions of years. Also, I should remind the reader that the reason for choosing DT for the first fusion reactors was its high probability of reaction at the lowest threshold temperature. All advanced-fuel reactors require much higher temperatures that may not be compatible with tokamaks (in which confinement appears to degrade as the temperature goes up) or with inertial fusion. While enthusiasts have periodically offered confinement schemes that they believe to be especially compatible with advanced fuels, such claims must be greeted with caution until more is known about them.

What is clear from the foregoing discussion is that fusion, though not yet a practical reality, deserves serious attention in a world where the need for energy and the environmental impact of energy production and use so clearly interact. A prudent energy development plan would include several options, as was intended in the plans of the 1970s, which have for the most part been aborted in the United States. These options might include the development of better fission systems and cleaner ways of using coal, the acceleration of research in fusion, solar power, and other "renewable" energy sources, and inducements to conserve. Our people and our leaders need to determine what kinds of governmental support are essential to make this happen.

Political leaders assume that the public has lost interest in energy since they rarely hear about energy from their constituents. I fear that this complacency about energy stems in part from a romantic view of the world, among liberals and conservatives alike. Some liberals believe that persuasion can alter people's energy consumption habits so greatly that conservation, not energy supply, is all we need think about. Some conservatives believe that, like Adam Smith's "unseen hand," the marketplace will wisely solve everything. Unfortunately, neither of these extreme positions is likely to be true.

We must assume that the energy demand of a billion people in China will somehow be satisfied regardless of environmental impacts, just as Italy's demand for electricity *did* continue to increase 5 percent annually even after its nuclear moratorium (and was only satisfied by buying nuclear-generated electric power from France!). In California, where I live, continuing battles against smog have prompted a law to promote electric automobiles. Yet, studies at the Pacific Gas and Electric Company have shown that although electric cars would indeed reduce pollution in Northern California, where much of our electricity comes from nuclear power or hydropower, in many Eastern states still dependent on electricity generated by coal, the wide use of electric cars at this time would only increase certain kinds of pollution by increasing the demand for coal-generated electricity.

At our 1990 Berkeley seminar on electricity supply and demand, I became involved in an interesting exchange when the group tried to focus on conclusions. Chauncey Starr, director emeritus of the Electric Power Research Institute, had set the stage. "What is the one big message that has come out of all this discussion?" Starr said. "The world is facing a chasm as far as electric power supply is concerned. And in varying degrees and different parts of the world, that chasm has various depths and various time scales about when you're going to hit the edge of it. And that chasm is so important it is going to override in importance around the world—including the U.S.—the whole issue of environmental concerns."

When another attendee expressed alarm, I observed that I had heard Starr's statement differently. "It has been interpreted as 'Energy *should* override environment.' I think what he is saying is that, unmanaged, it *will* override environmental protection actions. So that the issue is, if we don't do things in advance, the priority will not go to environment and overall good. The priority will go to meeting immediate needs. The whole point of management is to recognize those things that are going to overwhelm you, and be prepared in advance."

Starr and others agreed. H. M. Hubbard, of Resources for the Future, formerly director of the Solar Energy Research Institute, added, "That's a better environmental strategy anyway! My concern is that the environmentalists don't recognize it."

There are those who do. In the closing pages of his autobiography, completed after his death by his associate Mary Alinder, Ansel Adams, an early stalwart of the Sierra Club, writes:

> Many of my friends are shocked by my support for nuclear power. I feel that unless we stop polluting our atmosphere we will poison ourselves and be just as dead as the bomb would make us. Nuclear power is, of course, potentially dangerous if less than perfectly managed. If very carefully controlled in every respect, it should provide all the energy we need until fusion power is fully developed. The problem lies largely in the modern demand for energy, the wasteful use of power, and the finite

limit of natural resources. Hence, I favor what I consider to be the lesser of two evils and support nuclear power. I find that most opposition brings forth no constructive alternatives, no answers that provide for our future power needs.

Fusion power is the ideal solution to the energy problem, and its development should be given the highest priority. In May 1983 we were invited to visit the Lawrence Livermore Laboratories to see the great installations in magnetic and inertial fusion research. It was an enormously impressive experience, and, with the thought of what safe fusion power could contribute to the world's energy supply, I became more confident in the future. During my visit with President Reagan, I suggested that he take ten billion dollars from his defense program and apply it to a crash program for magnetic fusion development. Reagan raised an eyebrow to my temerity, but I believe it is obvious that once fusion is achieved, the energy shortage will be past and we will be independent of foreign fuels. In 1902 the automobile was in its infancy and the airplane an insubstantial dream. From the two-cylinder gas buggy to magnetic fusion is a giant stride, but, incredibly it can be accomplished during one lifetime.

16 : Predictable Breakthroughs

t was around 1980 that a noted plasma physicist prophetically quipped at a meeting, "We'll know we're getting somewhere when they stop asking whether fusion power will work and start asking what it will cost." The size and cost of the tokamak are now being questioned as we look past the great success in TFTR toward ITER. In anticipation of this, the ESECOM group mentioned in the previous chapter was asked to project the cost of electricity from magnetic fusion power plants.

The ESECOM group concluded that the cost of electricity produced by magnetic fusion reactors based on the tokamak would fall in the high end of the range of costs experienced by fission reactors, and that some of the more speculative designs considered in the study might compete with the lowest fission prices today. Since the best fission experience is currently competitive with electricity produced by coal, one would conclude that in terms of cost, although economic forecasts peering decades into the future are necessarily uncertain, magnetic fusion will probably be neither better nor worse than its competitors. These conclusions did not depend on any economic benefit accruing from the better Level of Safety Assurance assigned to fusion reactors. Similar conclusions are likely to apply to inertial fusion reactors, according to a recent status report issued under the auspices of the International Atomic Energy Agency of the United Nations.

Specifically, the ESECOM group compared the cost of electricity for the same four fission plants and ten conceptual magnetic fusion designs used in its environmental and safety comparisons, and also average costs for fission reactors operating today (see chapter 15). Because the fuel cost is low in all of the cases compared, the cost of electricity depends mainly on the cost of amortizing the capital investment in the plant. Setting aside the conventional equipment (e.g., steam turbines and generators) common to all cases, the greatest variation in cost is attributable to the "nuclear island"—the reactor itself and the heat delivery system associated with it. For the best of the fission reactors studied, the nuclear island represents only about one-third of the total plant cost, but in some existing fission plants and in projections for the fission breeder the cost of the nuclear island is several times as great, as it is for most of the fusion cases. Nonetheless, because the nuclear island is only a part of the cost, the cost of electricity varies much less. The actual results from the ESECOM study show a twofold variation in total capital cost for the fission cases, corresponding to a 70-percent spread in the cost of electricity. The fusion capital costs ranged from 1.2 to 2.2 times the lowest fission cost, with a cost of electricity ranging from 6 percent to 63 percent higher than the lowest fission case. On the basis of these findings, the ESECOM group's published paper concludes, "The cost of electricity for fusion reactors is comparable to that for fission reactors given the range of actual fission experience."

The tokamak reactor designs considered by the ESECOM group included anticipated improvements not yet incorporated in ITER. Some of these are plasma performance improvements that could have been tested in the Tokamak Physics Experiment recently proposed by the magnetic fusion community and now set aside because of budget reductions. True to the "dual path" development strategy described in chapter 14, concept improvement would have been pursued in TPX in parallel with advancements toward ignition and nuclear engineering in ITER. The combined results from TPX and ITER would have allowed one to design as a follow-on to ITER a Demonstration Power Plant embodying the best features of the tokamak identified to date.

The TPX-ITER development sequence would have continued the conservative evolutionary path of development that has been followed successfully by the U.S. and the international tokamak community for the last twenty-five years. The ESECOM study and other such studies lead one to conclude that this path of development could reasonably be expected to yield, by the mid-twenty-first century, fusion electric power plants that could replace a typical large fission power plant of today but would have greatly improved environmental and safety characteristics, producing electricity at a cost perhaps 50 percent above current best prices. Even a direct clone of ITER could do the same, but at a somewhat higher cost, resulting in rates perhaps double the current ones. Whether, fifty years from now, such prices for electricity will be considered high or a bargain is hard to say.

As I have tried to explain throughout this book, such claims can no longer be dismissed as wild-eyed dreams, though nothing is ever proved until it is done. But to make the dream come true, and to test the reality of this dream, society must be willing to sustain the high cost of completing the development of fusion, which is roughly, in constant dollars, the same as the U.S. level of commitment that was made in 1977 but not followed through. In other words, what is needed to make this happen is not a scientific breakthrough but a political breakthrough.

While I am in no position to predict political breakthroughs, in the closing pages of this book I would like to mention three scientific breakthroughs on the horizon that are in the best tradition of the "dual path," forging ahead with what we know while looking for something better. One such idea comes from inertial confinement fusion; the other two, closely related to each other, come from magnetic confinement fusion. Any one of these ideas could hasten the day when fusion energy becomes a reality. I could have mentioned other ideas, though the three I have chosen seem to me to offer the greatest promise at this time. However, as Edward Teller remarks in the preface of a 1981 monograph on fusion that, by chance, like this book, had sixteen chapters, "I cannot exclude the possibility that had we written 20 chapters, the additional four might have contained the final solution."

The new inertial fusion idea, called the fast ignitor, depends on new developments in laser technology. The ICF approaches discussed above use a single laser system that both compresses the pellet to a high density and heats the center to create a hot spot that ignites the fuel. When ignition begins, the pressure is roughly uniform throughout, implying that the burning hot spot at the center has a lower density than the surrounding colder shell. Since higher density reduces energy requirements, a higher yield would result if the hot spot were as dense as its surroundings. This is the goal of the fast ignitor concept.

This goal is to be accomplished in three steps, each of which, in this discussion, will be assumed to require its own laser, though in actuality the second and third steps could perhaps be combined and the total driver package may actually be smaller than that for the conventional approach. In the first step the pellet is compressed in the usual way, using a laser or any of the other drivers discussed in chapter 13. In the second step a smaller laser is used to punch a hole through the coronal atmosphere left over from the compression stage. And in the third step a small but very high-power "ignitor" laser, aimed down the hole, creates a hot spot on the surface of the dense, compressed fuel. Once this hot spot is ignited, "propagating burn" ignites and consumes the rest of the pellet fuel. According to calculations, this fast ignitor scheme greatly reduces the compression ratio required for a high yield, thereby reducing the size and complexity of the main "compressor" laser system. It could also increase the gain and yield in the NIF.

The technology development that makes the fast ignitor concept feasible is a new laser technique, called "chirped pulse amplification," that causes the ignitor laser pulse to compress in time so that its energy can be delivered more quickly — that is, at higher power. The ignitor laser deposits most of its energy in very energetic electrons that can penetrate into the core and initiate fusion burn more rapidly than the heat can leak into the outer fuel. By the time this book is published, Livermore expects to test a one-thousand-joule ignitor laser known as the "Petawatt."

The Petawatt laser is an example of a high payoff resulting from speculative technology development and basic research arising from a general interest in increasing laser power in order to study the basic science of laser-matter interaction in unexplored regimes. These high-power lasers are made possible by the technique of chirped-pulse amplification and a virtual revolution in optical materials and components that has occurred over the past few years. A chirped pulse is one in which the different wavelengths, or "colors," come at different times. Chirped-pulse amplification is a technique that was developed in the 1950s for overcoming the peak power limitations in radar systems. Stretching the radar pulse in time allowed more energy to be extracted from the radar tubes without causing damage to the optical components of the amplifiers. After the radar pulse left the amplifiers, it was compressed in time to increase its peak power.

It was recognized as early as 1975 that chirped-pulse amplification would bring about dramatic increases in laser power. The first practical demonstration of chirped-pulse amplification in lasers occurred at the University of Rochester in 1985. Using fiber optic techniques being developed for the communications industry, Donna Strickland and Girard Mourou extended the duration of a low-energy laser pulse to four times its original duration, amplified the pulse, and then recompressed it to a duration much shorter than the initial pulse. Initial propagation in the fiber created the bandwidth or "colors" necessary to create the short pulse upon compression.

This demonstration of laser chirped-pulse amplification occurred in the midst of extremely rapid developments in the design and use of ultrashort-pulse lasers and the optical materials necessary to produce them. By 1990, the original fiber optic technique for creating the chirped pulse was replaced by more direct methods utilizing "diffraction gratings" and new laser oscillators capable of generating pulses of such short duration that scientists had to learn a new word to describe them: "femtosecond," one second being a thousand trillion femtoseconds. Today, pulses as short as nine femtoseconds are produced. It is in the nature of light waves that such brief pulses cannot be waves of a pure color but instead must be a mix of many colors, many wavelengths. Propagation of such a pulse through a diffraction grating or prism separates the colors of the pulse. As a result, the pulse first stretches in time, since the red components travel a shorter distance than the blue. This "positively chirped pulse" (red coming earlier than blue) is then amplified in energy by as much as a trillionfold and finally recompressed to a duration close to the original pulse length. As of this writing, laser pulses one-eighth of the power of the upcoming Petawatt laser have been generated using this technique. The Petawatt laser itself, under construction in 1995, will produce pulses in excess of a thousand trillion watts (one "petawatt") for a very brief period of time.

The technology required to achieve petawatt pulses has itself produced numerous commercial benefits. Compression of the high-energy pulses following amplification required the development of new types of diffraction gratings manufactured at over one meter square. This technology is now being applied to the fabrication of flat panel displays (a multi-billion-dollar commercial market). Smaller-scale, chirped-pulse lasers are being used in medical and dental applications, with results that far surpass those achieved with conventional, longer-pulse lasers. These are applications far removed from the initial ICF application of the Petawatt laser, which demonstrates again the fact that we often cannot predict what benefits basic and applied research will yield.

The Petawatt laser development at Livermore has been carried out by a small group led by Michael Perry, a young nuclear engineer who received his doctoral degree from the Department of Nuclear Engineering at Berkeley in 1987. I first met Perry when he took my introductory course in fusion as a second-year graduate student in the fall semester in 1984. I realized that he was an exceptional student when he got a perfect score, thirty points ahead of the rest, on my first class assignment to "design a tokamak"—the engineer's version of the assignment we worked through back in chapters 2 and 3. I soon learned that he had been our top undergraduate senior in 1983 and held a nuclear science fellowship from the U.S. Department of Energy that encouraged his participation at Livermore for thesis research. As I was still a Livermore employee at that time, I encouraged Perry to join our magnetic fusion program. But by then, after an abortive attempt to tackle a forefront problem in neutrino basic physics, he had become interested in laser fusion.

Perry soon became a protégé of Michael Campbell, now head of laser research at Livermore. Campbell suggested that Perry look into another "basic science" problem, the ionization of atoms by light via "multiphoton" processes that are only possible with the very intense light produced by lasers. Meanwhile, in 1985, as a course assignment at Berkeley, Perry, who was now thinking of lasers full-time, wrote a paper on "pulse compression," which his professor thought to be a wonderful research topic. But Perry already had his topic. He did show the paper to Campbell, about six months before Strickland and Mourou published their breakthrough paper on "chirped pulse" amplification. Perry finished his thesis on multiphoton excitation in 1987 and immediately joined the research staff at Livermore, where, with Campbell and others, he pursued multiphoton excitation as a different approach to creating a very-short-wavelength "x-ray" laser. This new research soon led to the construction of a "chirped pulse" neodymium glass laser producing 10 million trillion-watt pulses, seven hundred femtoseconds in duration.

About this time, in part motivated by Perry's success, Campbell initiated a series of seminars at Livermore aimed at finding ways to improve inertial fusion reactors and reduce their cost. "He encouraged me to suggest outrageous concepts," recalls Max Tabak, who did offer ideas, one of which would become the fast ignitor. Some years earlier, long before the Petawatt laser was dreamed of, Tabak and others had pursued ideas along the lines of the fast ignitor, based on other technology. But now it would all come together, using the exciting new lasers that had initially been developed for basic science. When calculations with the LASNEX computer program confirmed Tabak's ideas, the fast ignitor was born; the basic concept was published by Tabak and colleagues in 1994.

The fast ignitor concept is a bold attempt to turn a problem into a solution. In chapter 12, we learned that the present success of ICF required a transition in technology from the red and infrared lasers of the early days to the green and ultraviolet lasers of today, mainly to prevent laser-plasma interactions that convert intense red light into highly energetic suprathermal electrons that can spoil compression of the fuel. Yet the fast ignitor depends entirely on the creation of suprathermal electrons that rapidly heat a hot spot at the surface of the fuel, which has already been compressed by other means. The compression phase is still better accomplished with a laser producing green or ultraviolet light, but thus far the Petawatt ignitor beam is red.

The success of the fast ignitor depends on understanding, measuring, calculating, and putting to good use some of the complex laser-plasma interaction phenomena that one largely hopes to avoid in the standard approach. The design issues, for the ignitor phase, turn on how deeply the ignitor beam penetrates through the dense ablation front before being absorbed, and how far the suprathermal electrons created by the beam penetrate into the fuel. If the beam stops too soon, in the low-density periphery, or if the suprathermal electrons' energies are so high that they propagate large distances in the dense interior, the suprathermal energy is spread over an excessively large volume and too much energy is required to heat the fuel to ignition. Like the standard approach, this is a complicated design problem, and at this stage of development of the concept it still involves many uncertainties.

If it all works out, however, the fast ignitor concept could greatly accelerate the development of inertial fusion energy and lead to a better product. The concept can be tested at high power by converting one of the ten Nova laser beamlines to amplify a chirped beam. If this test is successful, a beamline on the NIF could be similarly converted, and if all goes well, the fast ignitor scheme could ultimately increase the gain on the NIF about tenfold, thereby converting this ignition facility to a "high-gain" facility. Finally, if all these tests are successful, the fast ignitor concept could be applied to inertial fusion reactor design. The higher gain might allow the use of smaller reactors and might allow a reduction of efficiency requirements for the driver, to such an extent that glass lasers of various types might again become candidate drivers for a power reactor. In that sense, though revolutionary in terms of physics principles, the fast ignitor could prove to be conservatively evolutionary by allowing a more direct path of development from the pellet experiments of today to an energy-producing reactor. Time will tell.

The second new idea I want to discuss has an old history. Sometime in the late 1970s, while having a drink after a long day in Washington, a friend and I began to speculate about technologies that would make a difference for fusion. One, of course, was a smaller, cheaper driver to obtain high gain for inertial fusion—just what the fast ignitor concept could provide if the physics works out. The other would be a magnetic confinement scheme so simple that the plasma does everything, with a minimal number of external gadgets to make it work. We already knew what that idea would be. I mentioned the principle behind it, the self-organization of magnetic fields, back in chapter 7. The device based on this concept is called a spheromak.

We were building a spheromak at Livermore at the time, but only as a starting point for one of our "mirror improvement" ideas. When the idea was abandoned in favor of the tandem mirror, Edward Morse, at Berkeley, took advantage of a Department of Energy policy that allows universities to borrow equipment no longer in use and took the spheromak equipment to Berkeley's Department of Nuclear Engineering, long before I joined the department. When I later became chair of the department, I had to help convince the university to build a protective shed over Morse's spheromak so that he and his students could continue working on it while the laboratory that contained it underwent extensive demolition and construction work. Morse continued to build up his spheromak facility, first by obtaining powerful microwave tubes—military surplus equipment from a radar station in the Pacific—to heat the plasma, and later by acquiring a larger power system (capacitor bank) to operate the facility, obtained from Princeton when it had to abandon its own spheromak program as budgets contracted in the 1980s. Meanwhile, the Los Alamos fusion group had built and operated spheromaks, initially in collaboration with the Livermore mirror work, but by 1990 they, too, had to stop.

The first theoretical paper on the spheromak, by Rosenbluth and M. Bussac, appeared in 1979, and in 1980 a report was published on the first spheromak plasma created in the laboratory, at the University of Maryland in College Park. By the mid-1980s, Martin Peng and others at Oak Ridge were beginning to talk about a similar device called the spherical tokamak, the third of our three "predictable breakthroughs." The first experimental results on spherical tokamaks, from a facility in England, were published in 1992; by 1993 Thomas Jarboe, at the University of Washington, and Masayoshi Nagata, in Japan, were publishing papers on the results obtained using spheromaks converted to spherical tokamaks.

The English results on energy confinement in the spherical tokamak were encouraging, and by the summer of 1993 I had become convinced that the Los Alamos spheromak results, with the proper interpretation, were very encouraging also. In the summer of 1994, the Department of Energy sponsored a meeting in Oak Ridge at which the spherical tokamak and spheromak data were discussed and criticized, and several groups proposed future experiments, some that might explore both configurations in the same facility. The Department of Energy continues to express interest in such proposals, though no new device has been approved as yet.

The great interest in the spherical tokamak and the spheromak lies in the fact that these devices are much smaller than the standard tokamak, primarily because they involve fewer components. The size of the standard tokamak is determined by the requirements for magnets to create the strong toroidal field. The main point of the spherical tokamak is to create the toroidal field using a central conductor that is as small as possible and to form the plasma around this conductor as closely as possible. The result is a compact toroidal plasma in which the major and minor radii are almost the same. This is a radical departure from previous tokamaks but not different in principle, and the spherical tokamak is subject to the same theoretical principles and methods as the standard tokamak. However, the differences in technology and potential reactor configuration are profound.

To minimize the size of the central conductor, Peng and his colleagues abandoned superconductivity in order to increase the current per unit area as much as possible. The price for this is that a large amount of electric power is needed to drive the current, but this price is not necessarily unacceptable given the great reduction in reactor size made possible by this change. Peng also abandoned most of the shielding of the conductor so that it becomes vulnerable to neutron damage in a DT system. Thus, periodic maintenance may be needed to replace the conductor, which is then radioactive. Also, neutrons absorbed in the conductor do not contribute to tritium breeding and heat production in the blanket, which is totally outside. But the conductor is so small that few neutrons are lost. Despite these questions, the results of the English experiments, Peng's persistent leadership in pursuit of the theoretical issues and the attractive simplicity of the spherical tokamak have created much interest in this concept, as already noted. Predictions of performance for the spherical tokamak, like predictions for the fast ignitor, cannot be made as confidently as predictions for the standard tokamak, in this case because of uncertainties in determining the energy confinement time in a geometry greatly different from those in which the empirical scaling of confinement has been studied in the tokamak of today. Some empirical evidence suggests that at a fixed current, confinement degrades as the ratio of major and minor radii is reduced as Peng proposes, in which case much of the benefit would be offset by the requirement for very high currents to achieve ignition in a spherical tokamak. Again, time will tell.

The spheromak goes one step further than the spherical tokamak by abandoning the externally generated toroidal field altogether. The magnetic field "architecture" still looks like that of a tokamak, with nested flux surfaces constructed by field lines spiraling round and round, but now the field is produced totally by the current in the plasma. In chapter 7, we learned that the current-driven instability that is prevented by the toroidal field in a tokamak may not be a problem after all. Taylor's theory of self-organization of the magnetic field suggests that even without the external toroidal field the plasma would seek for itself a stable magnetic angle of repose. Moreover, also in chapter 7, I argued that, as my friends at Los Alamos had believed long ago, the "tearing" instabilities that accomplish self-organization should become progressively weaker as the plasma temperature increases, so that the leakage of heat by this process would become less than heat leakage via other mechanisms, such as the drift waves already dominant in the tokamak. Thus it may be that we need no longer fear the problem that caused us to introduce the toroidal field in the first place.

By the time spheromak experiments had ceased at Los Alamos in 1990, the self-organization of the field in spheromaks had been well demonstrated, but there continued to be doubt about the energy confinement time. A great improvement in energy confinement had resulted when Jarboe, then at Los Alamos, had suggested changing the "flux conserver" surrounding the plasma from an open wire cage to a smooth metal surface. At that time the experimenters were calculating the energy confinement time as the kinetic energy stored in the hot plasma core divided by the heating power due to the current. The magnetic heating power was taken to be the magnetic energy divided by the time the field lasted after all power sources were turned off. With the wire-cage flux conserver, the field duration was about the same whatever the plasma temperature, but with the new smooth flux conserver the field lasted longer when the plasma was hotter, as it should if the magnetic energy were being dissipated in the plasma due to the plasma's electrical resistance, which decreases as the temperature increases.

According to Jarboe, the problem with the cage was the tendency of self-organization to create current and field lines that intercepted the metal cage. Then the field would die there, and it was this death and regeneration at the cage that was using up most of the magnetic energy. To interpret all this wasted energy as heating of the plasma, as was being done to calculate the energy confinement time, was only confusing. In thinking about this, I concluded that even with the smooth flux conserver, most of the magnetic energy was still being wasted at the edge, where the plasma was cold and the resistance high, so that the Los Alamos interpretation of the energy confinement was still confusing. In fact, I estimated that the energy confinement time in the hot plasma core—which is what really matters—was ten or more times what was being quoted, and was probably in line with the heat leakage to be expected from the self-organizing process at the temperatures in the experiments, or alternatively, drift waves. After some discussion, Jarboe joined me and one of my students in a paper along these lines published in 1994. This has now led to an experiment, mentioned below, to see if we are right.

Like the ICF fast ignitor, the spherical tokamak and the spheromak could accelerate the development of fusion reactors and improve the product. But the magnetic fusion situation differs from the fast ignitor situation in several aspects. First, the potential improvement gained by making the reactor more compact must be weighed against the environmental and safety consequences of a higher power density, more fusion power, and more neutrons per unit volume. In the ESECOM study, while the design with the highest power density did give the lowest cost of electricity, it also had the poorest Level of Safety Assurance (level 3), because of the need for active cooling in an emergency. In this respect, inertial fusion may have an advantage because of the protective liquid or granular layer contemplated for inertial fusion reactors (see chapter 13). It may also be possible to incorporate liquid layers into high-power-density magnetic fusion reactors, but more study is needed before claiming this.

A second difference is the fact that, with glass lasers as drivers, the fast ignitor would be only an evolutionary step forward along the present course of development, whereas on the face of it the spheromak and spherical tokamak—at a stage of development comparable to the early tokamaks—are thirty years behind. However, just as a different kind of inertial driver could probably be developed in time to be a candidate for the stage beyond the NIF, given the smaller size of the spheromak and spherical tokamak and the great advances in theory, I believe that these concepts could also be developed in time to be candidates for the development stage beyond ITER.

It is their small size, requiring smaller and less costly development steps, that creates the peculiar appeal of the spheromak and spherical tokamak— all the more so if, as in the case of the spheromak, the number of auxiliary systems can be reduced to a minimum by letting the plasma do most of the work. As early as 1985, a reactor design published by R. L. Hagenson and Robert Krakowski demonstrated the potential virtues of a spheromak reactor. In its essentials, their reactor is just a cylindrical can perhaps six meters in diameter, like a big soup can, with circular coils to produce the "vertical" field, and the blanket wrapped around the can, and a "low-tech," low-voltage source of power to start it up. The plasma is produced by DT gas fed into the can. The power drives a current through the plasma, which, by self-organization, creates the spheromak current and magnetic field architecture inside the can, the current itself perhaps being able to heat the plasma to ignition. As has already been shown in experiments, a continuous current flowing down the axis will sustain this magnetic configuration in a steady state. According to Hagenson and Krakowski, at reactor temperatures the power consumed to maintain the field is less than 1 percent of the fusion thermal power produced. And there is a natural exhaust or "divertor" where the vertical field exits the can.

The relative simplicity of Hagenson and Krakowski's spheromak design suggests to me that one should think about how to package it as a component of an existing commercial power plant, just a replacement of the heat source. Developing a new component for an existing system is more financially feasible than undertaking the development of a whole new power plant. If this approach could indeed be applied to the spheromak, perhaps industry-government alliances to develop fusion energy would make sense sooner than we think. The spherical tokamak offers similar possibilities, and indeed, because the spheromak and the spherical tokamak are so similar, it only makes sense to develop them together, probably in a common facility, in order to learn directly the pros and cons of the engineering challenges posed by the central conductor versus the physics challenges of a self-organized plasma without the protective control of an external toroidal field.

Can such dreams come true? I am writing this on a Friday. On Monday I will drive out to Livermore to meet with two of my former graduate students so that the one who worked on spheromaks for his thesis can tell the other how to calculate for me the pressure limits for a spheromak, using an existing computer program. As we learned, in a magnetic fusion device it is the pressure and the energy confinement time that together make up the Lawson criterion of success. I am convinced that Peng's spherical tokamak will hold very high pressure, which is one of its advantages. I am convinced because that part of the theory of magnetic confinement is well understood now, and Peng and others have done the calculations. The spheromak experiments by Jarboe and his colleagues also produced an adequate pressure for a spheromak reactor, but I am unsure, looking back at older published results, how this fits theory. So we will calculate the pressure limit again, for the actual configurations we have in mind today.

I am especially anxious to know the pressure at this moment because I have an invitation to speak about the spheromak at an international meeting coming up in a couple of months. I would like to be able to say that I am even more convinced that the spheromak is a good bet. I know that some colleagues doubt my claim that the Los Alamos experiment demonstrated good heat confinement, but I will continue to be confident about that unless new experiments prove me wrong. Yet I have this nagging doubt about the pressure. In the old days, I would have simply gathered colleagues around, gone to work, and settled it—that is, assuming we had a credible theory and computer program for the question at hand. Now we have the computer program; but with fewer resources around, I have to be more patient.

Besides the pressure question, there is, of course, still the other half of the Lawson criterion, the energy confinement time, as well as the overall stability. Others are thinking about the stability, so I have set that aside for now. Probably some kind of "feedback" circuit will be needed, to push the plasma back in place if it tries to tilt or shift; and this may influence the practical choices between this added complexity for the spheromak and the toroidal field of the spherical tokamak. But these, I believe, are matters of competent engineering design.

The energy confinement time continues to be a more fundamental issue, despite all the theoretical progress reported in chapter 7. Later I hope that we can apply the wonderful new theoretical tools, the PIC computer simulations referred to in that chapter, to the spheromak, but for now I am more concerned about an experimental confirmation of my contention, mentioned above, that the spheromaks at Los Alamos confined heat much better than people thought—roughly as well as the T-3 tokamak in Moscow that started the tokamak revolution nearly thirty years ago. For the time being, since the Los Alamos spheromaks are no longer in operation, everything depends on the spheromak that Morse salvaged at Berkeley. A year ago, he was able to borrow the equipment to measure the temperature via Thomson scattering (see chapter 8). By heating the plasma core directly with his microwave tubes, Morse hopes to obtain an unambiguous measurement of the energy confinement time in the core and thus show directly whether or not the spheromak confines heat as well as I claim.

Of course, in the spheromak as in the T-3 tokamak, everything depends on getting an accurate measurement of the electron temperature with Thomson scattering. Fortunately, a doctoral student with a Department of Energy fusion fellowship (now scarce because of funding) took on the Thomson scattering measurement as his Ph.D. thesis project.

It would be wonderful to have an electron temperature measurement before my talk at the international meeting. The student has worked very hard, even to the point of stopping to tear down and rebuild the spheromak when this proved necessary to accommodate the laser and optical viewing equipment. He has everything back together now, despite an interruption to take his Ph.D. oral examination. Tomorrow, Saturday, I will attend his wedding, another essential if more pleasant interruption. Meanwhile, my estimates of the pressure limits look encouraging, and the electron temperature measurement, needed to determine the energy confinement time, will also yield an experimental determination of the pressure. I'll point this out to the young groom at the reception. He still has a couple of months to work before my talk.

Epilogue

O n May 22, 1996, the Petawatt laser at the Lawrence Livermore National Laboratory produced its first laser pulses exceeding the project goal of 1,000 trillion watts of power, thus setting in motion preparations for experiments on the Nova laser to test the exciting new fast ignitor concept, which could greatly accelerate the future development of inertial confinement fusion. On September 28, after further delays to improve the facility, the student working on the spheromak at Berkeley finally began to obtain the Thomson scattering measurements of electron temperature needed to test my optimistic interpretation of the energy confinement time in spheromaks. And throughout 1996, the major tokamaks around the world were demonstrating remarkable improvements in energy confinement time with "reversed shear." Through these exciting advances, and many other advances around the world, steady progress in fusion research goes on.

Yet, fusion's political dilemma continues. In the United States, the defense-related NIF for inertial fusion is being funded in 1996, while on the energy side the Congress has cut the research budget for magnetic confinement fusion, and also solar and renewable energy, by almost one-third below their 1995 levels, and more cuts are slated for 1997. Gone is the proposed TPX, and with it the hope to follow through with new facilities in the United States to exploit the exciting new discoveries of improved energy confinement times by "reversed shear." The future of ITER hangs in the balance, as Director Anne Davies refocuses the reduced U.S. magnetic fusion program on "fusion energy science."

Will the fate of fusion energy development matter? While there is no disputing that fusion is the most abundant energy resource available to future generations, barring a visible environmental collapse of global proportions one is hard pressed to prove the urgency of fusion as the solution to an immediate crisis. Yet it was not this that caused Secretary Gorbachev, President Mitterrand, President Reagan, and other world leaders of the 1980s to turn to fusion research as a worthy common goal toward which scientists could strive together, the spirit behind ITER. Will we, in the 1990s, be able to carry through? In the published scripts from his brilliant film series *Civilisation,* Kenneth Clark, a popular expositor of Western art and culture and former director of the National Gallery in London, equated civilization with, "above all, the confidence necessary to push through a long-term project." As such a project, fusion has not disappointed us. If the path to success has been longer than we once hoped, so also the science of fusion has proved richer than we dared to dream. May the dream live on.

Designing Your Own
. . . .
Fusion Reactor
. . . .

Now we will put numbers into the rules for designing a magnetic fusion reactor that we learned in part I of this book, so that the interested reader can calculate firsthand the magnetic field, current, and physical size required for a working reactor such as ITER. We will do a similar calculation for an inertial fusion reactor, applying the concepts developed in part III.

The starting point for both the magnetic confinement and inertial confinement approaches to fusion reactor development is the Lawson criterion. The reader should recall that the Lawson criterion for magnetic fusion, which indicates whether the plasma will be able to ignite and burn by itself, is a threshold value of a certain number obtained by multiplying the plasma pressure by the energy confinement time. We have called this the Lawson number, after the English physicist J. D. Lawson, who first published the criterion for fusion ignition with a mixture of deuterium and tritium (DT) fuel back in 1956. I should note in passing that in Lawson's paper and in the magnetic fusion literature, the Lawson number is defined as the energy confinement time multiplied by the particle density (ions per cubic meter) rather than the plasma pressure. Then the Lawson criterion depends on temperature, whereas, as we shall see, the temperature dependence has already been included in the Lawson number itself as I have defined it. The Lawson number is used somewhat differently in ICF.

For magnetic fusion, the Lawson criterion for ignition can be stated quite precisely. The idea, remember, is that the fusion power itself heats the fuel to the temperatures required for fusion reactions to occur, despite the leakage of heat by radiation and by conduction. Barring impurities, at a temperature of 100 million degrees Celsius the loss of heat by radiation is small and can be neglected. Then the heat leakage is just the heat energy due to the plasma temperature divided by the energy confinement time due to heat conduction. Only the alpha particles heat the plasma, since the plasma is transparent to the energetic neutrons produced by DT reactions. Ideally, all of the alpha energy serves to heat the plasma, so the heating power is simply that portion of the fusion power (one-fifth of the total) produced as energetic (fast-moving) alphas. The fusion power can be calculated from the fuel density and a characteristic DT reaction rate that physicists call the "cross section," which has been carefully measured in the laboratory.

Putting this together, we find as the condition for ignition

$$\left(\frac{n}{2}\right)\left(\frac{n}{2}\right)(\sigma v)(5.6 \times 10^{-19}\,\text{MJ}) = \left(\frac{3nT}{\tau}\right)(1.6 \times 10^{-29}\,\text{MJ}). \tag{1}$$

The right-hand side of this equation gives the energy leakage in terms of the energy confinement time, given the symbol τ (in seconds), the fuel density, n (in ions per cubic meter), and the temperature, T (in degrees Celsius). (More properly, the temperature should be in degrees Kelvin, measured from absolute zero, but the difference between the Kelvin and Celsius scales— 273.2 degrees—is negligible at fusion temperatures.) The factor 3 relates temperature to the average energy per pair of ions and electrons, and the factor $(1.6 \times 10^{-29}\,\text{MJ})$ converts the answer to megajoules (MJ), or millions of joules, the joule being a unit of energy (4.19 joules equals 1 calorie, the heat energy needed to increase the temperature of 1 gram of water by 1 degree Celsius). Here we are using scientific notation, in which "10^{-29}" means "move the decimal point twenty-nine places to the left," and "10^{29}" means "move the decimal twenty-nine places to the right." Thus, for example, 10^{-6} means "divide by 1 million," while 10^{6} means "multiply by 1 million," and so on.

The left-hand side of equation (1) is the alpha heating power per unit volume, in terms of the cross section, σ (in square meters); the ion velocity, v (in meters per second); and the fuel density, n. The fuel density appears twice to account for a reaction of deuterium (half the fuel, hence the factor 1/2) and tritium (also 1/2). The quantity $5.6 \times 10^{-19}\,\text{MJ}$ is the energy of an alpha particle produced by a DT fusion reaction. The product σv gives the DT reaction rate. As mentioned above, we are assuming an ideal situation in which none of the alpha energy escapes before collisions with fuel particles transfer the alpha particles' energy to the fuel. This is probably a good approximation of an actual situation, according to the results from the TFTR experiment discussed in chapter 4.

Having thus found all the pieces required to calculate the condition for ignition, we can now solve equation (1) to obtain the Lawson number. In mathematical terms, the Lawson number as I have defined it is $nT\tau$, meaning "the fuel density multiplied by the temperature, to get the pressure, and multiplied again by the energy confinement time." Then, moving quantities around as one would in a high-school algebra problem, we find

$$nT\tau = (2 \times 2 \times 3)\left[\frac{(1.6 \times 10^{-29})}{(5.6 \times 10^{-19})}\right]\left(\frac{T^2}{\sigma v}\right). \tag{2}$$

Notice that now the density appears just once—in equation (1) it appears once on the right and twice on the left—and we have multiplied each side by the temperature to get the desired form for the Lawson number.

At a temperature of 100 million degrees Celsius, $\sigma v = 1.1 \times 10^{-22}$ in units of cubic meters per second. Then the Lawson criterion, given by equation (2), becomes

$$nT\tau = 3.1 \times 10^{28} \tag{3}$$

in units of degrees Celsius times seconds per cubic meter. This is the Lawson criterion, which determines whether steady-state ignition can be achieved in a magnetic fusion reactor. Though derived here for a temperature of 100 million degrees, equation (3) is valid in the range of 100 to 200 million degrees Celsius, since $T^2/\sigma v$, appearing in equation (2), is approximately constant in this range.

We will now apply the magnetic Lawson criterion to design the ITER tokamak, as we did conceptually in chapters 2 and 3. In chapter 3, we learned that experiments in many tokamaks have suggested an "empirical scaling law," the Goldston scaling, which simply says that the Lawson number is proportional to the square of the current in a tokamak. When numbers are added, this scaling law becomes

$$nT\tau = I^2 \times (7.0 \times 10^{25}), \tag{4}$$

where the numerical factor is chosen to fit the experiments, with the Lawson number expressed in the same units as in equation (3). Actually, this numerical factor varies somewhat from one experiment to the next. Here I have chosen the value that reproduces the value of the current in the present ITER design, based on the best judgment of the designers given the experimental data available today, and I have absorbed any dependence on geometric factors in this numerical factor. The units are chosen so that the current, I, is given in megamperes (MA), or millions of amperes. (The currents in household appliances are typically only a few amperes.) Then, substituting the Lawson criterion value on the left in equation (4) tells us that $I = 21$ MA is required for ignition in ITER, in agreement with the table on page 124. In other forms of the empirical scaling law that also fit the data, the current may appear to a different power (e.g., $I^{3/2}$).

Knowing the required current, we can now work backwards to find the size of the machine and the fusion power it produces, using the pressure rule and the current rule obtained from the energy principle in chapter 2. There, we found that in order to avoid kinking instability the current must not exceed a value found by multiplying the field strength times the plasma minor radius. To reduce this to actual numbers requires working out the geometry of magnetic field lines in the tokamak to ensure that no field line twists more than once around the magnetic axis in traveling once around the torus ($q \geq 1$). The answer, for a plasma somewhat elliptical in shape as in ITER, is that the current I must not exceed the value given by the following equation:

$$ I \leq \left(\frac{aB}{A} \right) \times 4.0. \tag{5} $$

where the symbol "$\leq$" means "no greater than." Here we have included the effect of the aspect ratio, $A = R/a$, mentioned in chapter 5. The numerical coefficient is appropriate for typical current profiles in which the "safety factor" q has the value 1 at the magnetic axis, increasing to 3 near the edge of the plasma.

In chapter 2 we learned that it is best to make the magnetic field strength as high as possible, 5.7 tesla (57,000 gauss) being the ITER design value, given in the table on page 124. If the field strength $B = 5.7$ tesla, the aspect ratio $A = 3$, and the current $I = 21$ MA, as required for ignition, this gives a minimum minor radius

$$ a = 2.8 \text{ meters.} \tag{6} $$

Allowing room for the magnets and blanket requires approximately a major radius $R = 2.9a$, or

$$ R = 8.1 \text{ meters,} \tag{7} $$

as given in the table on page 124. Approximating the plasma volume as an elliptical cylinder with a ratio of major and minor axes of 1.75, we find the volume, V, to be

$$V = 2 \times 1.75 \times \pi^2 \left(a^2\right) R = 2{,}190 \text{ cubic meters,} \tag{8}$$

where $\pi = 3.1416$. The overall height of the machine, roughly $5 \times R$, is about 40 meters, or 130 feet—big, but not outside the scope of large industrial enterprises.

Finally, we can obtain the fusion power, P, by multiplying the volume times the alpha heating power per unit volume, given by the left side of equation (1), and multiplying the result by a factor of 5, to take into account the energy in neutrons. Thus

$$P = 5\left(\frac{n}{2}\right)\left(\frac{n}{2}\right)(\sigma v) \times V \times (5.6 \times 10^{-19} \text{ MJ}), \tag{9}$$

where $\sigma v = 1.1 \times 10^{-22}$ as before. To complete the calculation, we need the density, n, which we can obtain from the pressure rule derived by Troyon and co-workers from the energy principle and shown to agree well with experiments. In chapter 2, we found that the pressure rule states that the pressure must not exceed a number found by multiplying the current times the field and dividing by the plasma minor radius; or, mathematically,

$$2nT \leq \left[\frac{(I \times B)}{a}\right] \times (4.4 \times 10^{26}), \tag{10}$$

where $2nT$ is the sum of the ion pressure *(nT)* and the electron pressure (also nT), with n in units of ions per cubic meter and T in degrees Celsius. Substituting $T = 10^8$, $I = 21$ MA, $B = 5.7$ tesla, and $a = 2.8$ meters gives

$$n = .94 \times 10^{20} \text{ per cubic meter,} \tag{11}$$

and substituting this into equation (9), with $V = 2{,}190$ cubic meters, gives approximately

$$P = 1{,}500 \text{ MW,} \tag{12}$$

as shown in the table on page 124.

This completes our calculation of the main parameters of ITER, as presented in chapter 10. The physics behind the design is incorporated into the various numerical factors given in these formulas. The reader can test the sensitivity of the design to uncertainties in the physics by changing these factors and then calculating the consequences. For example, suppose that the experimentally determined factor in equation (4) was too high, requiring us to recalculate the current and therefore the minor radius and the fusion power derived from this current. Following through the chain of reasoning, we would find that a 10 percent decrease in the numerical factor in equation (4) would yield a 5 percent increase in the current, bringing it to 22 MA, and hence a 5 percent increase in the minor radius, to $a = 2.9$ meters. R also increases, to 8.5 meters, and the volume thus increases to $V = 2,465$ cubic meters, while the density remains the same, since I and a change together in equation (10), yielding a 13 percent increase in fusion power, to 1,700 megawatts (MW). Conversely, a smaller numerical factor in equation (4) would permit a somewhat smaller machine.

We can also use the above formulas to see the potential benefits from the spherical tokamak "breakthrough" idea discussed in chapter 16. An important difference is that R and a are nearly the same, perhaps as close as $R = 1.1a$ (rather than $3a$, as in equation [7]). This reduces the volume by a factor of $(1.1/3.0) = 0.37$, other things being equal. It also reduces the aspect ratio, A, appearing in the current rule given in equation (5), thereby allowing more current in a smaller machine. Then, if the Lawson number still scales like equation (4) so that the same current is required for ignition (a big "if"!), the required minor radius might be one-half or one-third as large, yielding a smaller machine. For example, if the minor radius turned out to be $a = 1.5$ meters, the plasma volume would be only

$$V = 2,190 \left(\frac{1.1}{3.0}\right)\left(\frac{1.5}{2.8}\right)^3 = 123 \text{ cubic meters.} \tag{13}$$

However, according to equation (10), at the same magnetic field strength the density would be higher by a factor of $(2.8/1.5)$, or $n = 1.8 \times 10^{20}$ per cubic meter, yielding a higher power density by a factor of $(1.8)^2 = 3.2$. Taken together, the fusion power would be

$$P = 1,500 \left(\frac{123}{2,190}\right)(1.8)^2 = 273 \text{ MW.} \tag{14}$$

Then ignition could be achieved in a machine of much smaller size and lower fusion power. However, as noted in chapter 16, these apparent benefits depend on uncertainties in geometric details that are omitted in equation (4).

Next, we turn to the design of an inertial fusion device capable of reaching ignition, such as the NIF. For ICF, performance is described in terms of the energy gain, G, defined as the energy yield in ratio to the laser energy. The onset of ignition is taken to be a gain of around 1, the nominal design point for the NIF being $G = 10$ to leave a margin for uncertainties. The gain is calculated from

$$G = f(3.5 \times 10^5 \, M/E_{laser}), \tag{15}$$

where M is the fuel mass in grams, E_{laser} is the laser energy in megajoules, and f is the "burnup fraction" given by

$$f = \frac{L}{(L + 7)}. \tag{16}$$

Here $L = \rho R$ is the Lawson number for the compressed fuel, defined — as is customary in the ICF literature — as the density, ρ, in grams per cubic centimeter multiplied by the radius, R, in centimeters. Dividing R by the average speed of a deuteron at 100 million degrees Celsius (around 10^8 centimeters/second) gives the burn time, which is also the energy confinement time. The fuel mass M is related to L by

$$M = \left(\frac{4\pi}{3}\right)\rho\left(\frac{L}{\rho}\right)^3 = \left(\frac{4\pi}{3}\right)\left(\frac{L^3}{\rho^2}\right), \tag{17}$$

where we use $(L/\rho) = R$ to compute the radius. The compressed fuel density is $\rho = 0.2C$, where C is the compression ratio, the ratio of fuel volume before and after compression. Introducing this in equation (17) and solving for L gives

$$L = (0.01MC^2)^{1/3}. \tag{18}$$

This expression can then be used in equation (16) to compute the burnup fraction in terms of M and C.

There are two contributions to the energy, one to compress the fuel and one to heat the hot spot. Then

$$E_{laser} = \eta^{-1}(E_{compression} + E_{hot\ spot}), \qquad (19)$$

where for a direct drive system the quantity η is the "hydrodynamic" (i.e., ablation) efficiency multiplied by the absorption efficiency; for indirect drive η is the hydrodynamic efficiency multiplied by the hohlraum efficiency.

The compression energy can be expressed in terms of the fuel mass, M, and the Fermi energy, which increases with compression. Again using $\rho = 0.2C$, the required energy, in megajoules, is approximately

$$E_{compression} = (0.25)\,MC^{2/3}\ \text{MJ}. \qquad (20)$$

This result assumes that pulse shaping is employed to avoid heating of the fuel as it is compressed (see chapter 11).

The hot spot energy is more difficult to calculate from simple models, but a rough estimate can again be obtained from an appropriate Lawson number, $L_{hot\ spot}$, for the hot spot, determined by competing factors such as the rate of hydrodynamic expansion (roughly the L for the main fuel); the rate of heat conduction from the hot spot into the main fuel; radiation from the hot spot; and finally the requirement that the alpha particles produced by DT fusion reactions deposit most of their energy within the hot spot so as to continue heating the hot spot. Whereas in magnetic fusion alphas are well confined and heat conduction by plasma turbulence dominates, in ICF heat conduction in the fuel (mainly collisional) diminishes at fusion temperatures to the point that requirements determined by conduction become comparable to the requirement that the hot spot radius exceed the range of alpha particles slowed down by collisions with electrons. This gives $L_{hot\ spot}$ around 0.3, only one-tenth of the value required for a high burnup fraction and high gain, in equation (16). By chance, dividing $L_{hot\ spot}$ by 10^8 centimeters/ second gives a Lawson number comparable to the magnetic Lawson number for ignition, while that for high gain is ten times as high. For magnetic fusion, in which fuel is supplied continuously, "ignition" also implies high gain, since the plasma burns steadily once ignited, whereas for ICF achieving high gain by prolonging the burn requires a correspondingly higher Lawson number.

Given the hot spot Lawson number, we can calculate the hot spot mass from equation (17), now interpreting all quantities as those appropriate for the hot spot. Introducing $L = 0.3$ (for the hot spot) into equation (17) gives the mass of the hot spot. To obtain the hot spot energy, we multiply this mass by the 600 MJ per gram needed to heat this mass to temperatures somewhat above the 50 million degrees Celsius required for alpha heating to exceed radiation losses. Then

$$E_{\text{hot spot}} = 600 M_{\text{hot spot}} = \left(\frac{70}{\rho_{\text{hot spot}}^{2}} \right) \text{MJ}, \tag{21}$$

where $\rho_{\text{hot spot}}$ is the density in the hot spot.

We first make the conservative assumption, mentioned in chapter 16, that the pressure in the hot spot is the same as the pressure required to compress the fuel. This pressure is approximately the compression energy we just worked out in equation (20) divided by the volume after compression. Dividing the fuel mass M that appears in equation (20) by the volume just gives the fuel density, $0.2C$, so the pressure is proportional to $C^{5/3}$. Then the hot spot density $\rho_{\text{hot spot}}$, which is just the pressure divided by 400 MJ per gram ($2/3$ of the heat energy at 50 million degrees Celsius) is also proportional to $C^{5/3}$, and $\rho_{\text{hot spot}}^{2}$ appearing in equation (21) is proportional to $C^{10/13}$. Putting in the numbers, we find that, with a uniform pressure, the energy required to heat the hot spot is approximately

$$E_{\text{hot spot}} = \left(5 \times \frac{10^{9}}{C^{10/3}} \right) \text{MJ}. \tag{22}$$

We can now apply these results to the design of the NIF, with 1.8 MJ of laser energy. Assuming a hydrodynamic efficiency of 15 percent and a hohlraum efficiency of 15 percent (2.2% overall), up to 0.04 MJ could be deposited to compress the pellet and heat the hot spot. According to our approximate formula, at a compression ratio of, say, $C = 3,000$, the energy to ignite the hot spot would be (using equation [22]) about 0.013 MJ, leaving 0.027 MJ to compress the fuel. According to equation (20), with $C = 3,000$ this would be sufficient to compress about 0.5 milligrams of mass. Then, from equations (16) and (18), we find the burn fraction $f = 0.34$, and from equation (15), the gain $G = 33$, which exceeds the nominal goal for the NIF.

The interested reader may wish to apply these simple formulas to construct curves depicting gain versus compression, along the lines of our discussion in chapter 11, with a different gain curve for different values of the laser energy. To do so, first choose a laser energy, E_{laser}, and an efficiency, η. Now substitute these values into equation (19), and also substitute into this equation $E_{\text{hot spot}}$, obtained from equation (22), and $E_{\text{compression}}$, from equation (20). With these substitutions, equation (19) determines M in terms of C for the particular laser energy used, and given M and C, the gain can be calculated for this value of C as we did above. Choosing a new value of C and repeating the procedure gives the gain for this new C, and so on, until we find the complete graph, or "gain curve," for the laser energy we have chosen. Because we have neglected the hot spot mass in our gain formula, the procedure fails if C is too small relative to the chosen laser energy (M would be negative). Then the gain curve for that laser energy begins, with zero gain, at the compression ratio that satisfies equation (22) for $E_{\text{hot spot}} = \eta E_{\text{laser}}$. For the same laser energy, the gain will increase as we increase C, because the contribution from the hot spot decreases. Finally a maximum gain will be reached, as the burnup fraction increases, after which a higher compression ratio only reduces the gain, since more energy is required to compress the fuel with no more relative yield. Just this behavior can be seen in the gain curves calculated by John Nuckolls and his colleagues in their path-breaking paper on ICF published in *Nature* in 1972, though of course, as in our ITER design, our simplified formulas only give a rough idea of the actual performance predicted by detailed analysis and computation. In particular, our formula for the hot spot energy exaggerates the benefits of compression at the ignition threshold ($G = 1$).

We can also apply our approximate formulas to assess the benefits of the fast ignitor breakthrough for ICF discussed in chapter 16. The main benefit comes from a reduction in the energy needed to heat the hot spot. For the fast ignitor concept, heating is so rapid that the hot spot density can be the same as that of the compressed fuel, or $\rho_{\text{hot spot}} = 0.2C$. Substituting this value into equation (21) gives

$$E_{\text{hot spot, fast ignitor}} = \left(\frac{1{,}750}{C^2}\right) \text{MJ}. \tag{23}$$

Then, $C = 3{,}000$ corresponds to an energy of only 2×10^{-4} MJ, or 200 joules, one-seventieth the hot spot energy required for uniform pressure. Thus, with the fast ignitor, one could afford to design for a lower compression ratio, requiring less compression energy for a given mass of fuel.

Glossary

Ablation. The process of compressing a pellet by intense heating of the surface, creating an exploding plasma that pushes inward on the surface as the plasma expands outward.

Active safety systems. In nuclear reactors, devices, such as emergency cooling systems, that must be activated if an accident occurs.

Adiabatic invariant. As used here, a theoretical quantity, such as the angular momentum of rotation (the speed multiplied by the gyroradius), that is almost constant during the course of motion of electrons or ions in a magnetic fusion device.

Advanced fuels. Fusion fuels other than DT, usually producing less energy in the form of neutrons.

Alcator. A high-field tokamak at the Massachusetts Institute of Technology.

Alpha particles. Helium nuclei, a product of DT fusion reactions. Also called "alphas."

Alternative concept. Any magnetic confinement device other than a tokamak.

Ampere. The common unit of electric current, sometimes shortened to "amp."

Amplifier. For glass lasers, a large chunk of glass that stores energy that later amplifies a light beam passing through it. A plasma column energized by electron beams serves this function in gas lasers.

Argus. An early laser facility at the Lawrence Livermore National Laboratory.

Atoms. The building blocks of matter. An atom consists of a positive nucleus containing most of the mass, surrounded by orbiting electrons.

Axially Symmetric Divertor Experiment (ASDEX). The tokamak in Germany where the H-mode was discovered.

Beamlet. A prototype beamline for the NIF.

Beamline. In ICF, the chain of optical components needed to generate and amplify a laser beam.

Binary collision. The collision of electrons and ions two at a time.

Blanket. A solid structure surrounding a fusion reactor, used to absorb neutron energy as heat and to "breed" tritium by neutron reactions with lithium.

Bootstrap current. Electric current self-generated inside the plasma in a magnetic confinement device.

Break-even. The condition in which fusion energy output equals the energy required to heat the fuel.

Cavity. In ICF, the inside of a hohlraum.

Centi-. A prefix meaning "one hundredth."

Chirped-pulse. A technique for compressing the duration of a laser pulse to increase the power.

Cold fusion. The popular name for claims, not accepted by most scientists, that certain chemical devices stimulate fusion reactions at ordinary temperatures.

Collective effects. A distinguishing property of plasmas in which large numbers of electrons and ions act together to create wavelike motions in a plasma.

Conductivity. The ability to conduct electricity; plasmas are conductors.

Corona. In ICF, the hot plasma surrounding a pellet during ablation.

Critical density. The density at which a plasma strongly reflects even low-intensity light.

Curie. A unit of radioactivity equal to 3.7×10^{10} nuclear disintegrations per second.

Current rule. As used here, an approximate formula used in designing tokamaks, relating the plasma current to the magnetic field strength and plasma dimensions.

Degrees Celsius. A temperature scale with zero at the freezing point of water. One degree Celsius equals 1.8 degrees Fahrenheit. Absolute zero is minus 273.2 degrees Celsius.

Deuterium. An isotope of hydrogen to be used as fuel in fusion reactors.

Deuterium-tritium fuel. (DT) fuel. A mixture (usually equal parts) of deuterium and tritium.

Deuterium-tritium (DT) reactions. Nuclear fusion reactions in DT fuel.

Diamagnetism. In magnetically confined plasmas, a property of the motion of ions and electrons in a magnetic field whereby a gradient in plasma pressure automatically produces the plasma currents required for the field to confine the pressure.

DIII-D. The second largest tokamak in the United States, at General Atomics, in San Diego.

Direct drive. The ICF approach in which laser beams coming from all directions illuminate the pellet directly.

Drift wave. A type of collective phenomenon in magnetically confined plasmas, associated with the turbulent transport of heat out of the plasma.

Driver. A means to ignite the pellet in ICF—for example, lasers or ion beams.

Early fatality. As used in fusion and fission reactor safety analysis, death by exposure to at least 200 rem of radiation, producing recognizable symptoms that can cause death, typically within sixty days after exposure.

Electromagnetic force. The combined effects of electricity and magnetism.

Electron. An elementary particle, negatively charged. All atoms consist of a nucleus surrounded by electrons.

Electron cyclotron heating. A method for heating plasma using intense microwave radiation.

Energy confinement time. The time required for most of the heat to leak out of a hot plasma; the cooling time. Specifically, the time needed for the heat to decrease to $1/e$ times its original value ($e = 2.718$, the base for natural logarithms).

Energy principle. A theoretical formula for deciding whether or not a plasma is stably confined according to MHD theory.

Entropy. A number related to the state of disorder of a physical system. For example, an increase in entropy occurs if directed motion such as an electric current is converted to heat (the process of electrical resistance).

Equilibrium. A state of exact balance of forces, as when the magnetic force exactly balances the outward pressure of a plasma in a tokamak.

ESECOM. The Senior Committee on Environmental, Safety, and Economic Aspects of Magnetic Fusion Energy, a panel of experts convened by the U.S. Department of Energy.

Fast ignitor. A new approach to ICF, using ultra-high-power lasers to create the hot spot after the plasma is compressed by other means.

Fermi energy. The quantum-mechanical internal energy of atoms, which increases when atoms are compressed in ICF pellet implosions.

Field line. An imaginary line denoting the direction of a magnetic field. In a magnetically confined plasma, electron and ion motion tends to follow field lines.

Filamentation. Self-focusing of light beams in glass or plasma causing the beam intensity to increase locally.

Fission. A nuclear reaction in which a heavy nucleus such as uranium-235 splits in two, with the release of energy and radiation.

Flashlamp. For glass lasers, the source of light that energizes the amplifiers.

Flibe. A mixture of lithium, beryllium, and fluorine.

Flux surface. An imaginary surface in space mapped out by magnetic field lines inside a tokamak or other plasma confinement device.

Free energy. That portion of the plasma energy available to create turbulence and increase entropy; related to the existence of current and the confinement of pressure in a tokamak.

Frequency. For light, the rate of change in time of the oscillating electric field associated with a light wave.

Frequency conversion. The process for converting red light to green or ultraviolet light.

Fusion. The joining of two atomic nuclei to form a heavier element.

Gain. In ICF, the ratio of the fusion energy produced to the driver energy absorbed by the pellet.

Gauss. A unit of magnetic field strength. The Earth's magnetic field is about one-third gauss at the equator.

Gekko II. A ten-kilojoule laser facility at Osaka, Japan.

Gram. A unit of mass; 454 grams equal one pound.

Gyroradius. The radius of a spiraling orbit of an electron or ion moving in a magnetic field.

H-mode, or high mode. An improved way of operating a tokamak to increase its plasma energy confinement time.

Halite/Centurion. A series of experiments with ICF targets using energy produced by underground nuclear explosions.

Heavy ion driver. An ICF driver using a particle accelerator to create intense beams of cesium ions, or other ions of heavy elements.

High Yield Lithium Injected Fusion Energy (HYLIFE). A design for an ICF reactor utilizing a protective layer of liquid lithium or flibe as the blanket.

Hohlraum. In indirect-drive ICF, a tiny can or cavity containing the pellet.

Hot spot. A tiny hot region, within a compressed pellet, that serves as a "fuse" to ignite the main fuel.

Ignition. The condition in which fusion reactions themselves heat the fuel sufficiently to sustain the reaction.

Implosion. In ICF, the inward burst of the pellet created by the inward acceleration of pellet mass during ablation.

Indirect drive. The ICF approach employing a hohlraum to create a symmetric implosion of pellets using x-rays generated by heating the walls of the hohlraum cavity using lasers or ion beams.

Induction. The mechanism whereby a magnetic field changing in time causes electric current to flow in a conductor.

Inertial confinement fusion (ICF). Fusion energy production by exploding tiny fuel pellets.

Instability. Any tendency of small disturbances of plasma equilibrium to have big effects, creating turbulence or even the rapid loss of confinement of the plasma.

Intensity. For lasers, the beam brightness or power per unit area.

International Atomic Energy Agency (IAEA). An agency of the United Nations that sponsors world conferences on fusion.

International Thermonuclear Experimental Reactor (ITER). A 1,500-megawatt experimental magnetic fusion reactor being designed as a joint project of the European Union, Japan, Russia, and the United States.

Intruder dose. A measure of the radiation exposure of an individual who might accidentally come in contact with buried radioactive waste.

Ion. An electrically charged atom. Positive ions have lost electrons; negative ions have gained electrons.

Ion beams. An alternative to lasers as drivers for ICF.

Ion cyclotron resonance heating. A method for heating plasma using high-power radio waves.

Japan Tokamak 60 (JT-60). Japan's largest tokamak, at Naka.

Joint European Torus (JET). Europe's largest tokamak, at Culham, England.

Joule. A unit of energy; 4.19 joules of energy, equal to one calorie, would increase the temperature of one gram of water one degree Celsius.

KDP. *See* Potassium dihydrogen phosphate.

Kilo-. A prefix meaning "one thousand."

Krypton fluoride (KrF) laser. A gas laser demonstrated to produce 0.25-micron light beams highly uniform in intensity, more efficiently than today's glass lasers.

Langmuir probe. An instrument developed by Irving Langmuir for use in measuring electric potential and other properties in plasmas.

Large Coil Test Facility. A superconducting magnet test facility at the Oak Ridge National Laboratory.

Laser. The generic name for devices producing very intense, highly focused beams of light.

Laser-plasma interactions. Collective plasma wave generation by intense laser beams.

LASNEX. A computer program developed at the Lawrence Livermore National Laboratory to simulate ICF pellet implosions.

Lawrence Award. An annual award by the U.S. Department of Energy recognizing outstanding achievements by individuals.

Lawson criterion. The condition that must be met to ignite DT fuel.

Level of Safety Assurance. A safety criterion developed to assess the relative safety of fusion and fission reactor designs.

Light ion driver. An ICF driver approach utilizing inexpensive technology to accelerate beams of lithium ions, or other ions of light (low-mass) elements.

Magnetic confinement fusion. Fusion energy production by a plasma confined by a magnetic field.

Magnetic island. The twisting of field lines about a central line to form local flux surfaces inside a tokamak or other confinement device.

Magnetic mirror. As used here, a magnetic confinement device using the property whereby the diamagnetism of a plasma causes it to be reflected when moving toward a region of higher magnetic field strength.

Magnetic well. An arrangement of magnets creating a strong magnetic field surrounding a location of minimum field strength.

Magnetohydrodynamics (MHD). An approximate plasma theory in which the plasma is assumed to behave as a fluid with infinite conductivity.

Main fuel. In an ICF pellet, the compressed fuel surrounding the hot spot.

Major and minor radii. In a tokamak, the plasma dimensions, major radius R and minor radius a (see fig. 1).

Master oscillator. In a laser, the generator of the trigger beam.

Maxwell Prize. An award by the American Physical Society to recognize outstanding contributions in plasma physics.

Mega-. A prefix meaning "1 million."

Meter. The metric unit of length, equal to 3.28 feet.

Micron. One millionth of a meter.

Milli-. A prefix meaning "one thousandth."

Mirror Fusion Test Facility (MFTF). The largest magnetic mirror device ever constructed, at the Lawrence Livermore National Laboratory.

Mixing. In ICF, the unstable mixing of cold fuel into the hot spot during an implosion.

Mixing length. The dimension over which motion is coherent within a generally turbulent plasma, a scale length used to estimate the rate of heat transport due to the turbulence.

National Ignition Facility (NIF). A large laser facility being designed to achieve ignition of DT targets.

Neodymium glass. The glass most often used to construct high-power glass lasers, containing the element neodymium.

Neutral beams. Beams of energetic neutral atoms injected into a plasma to heat it.

Neutron. An elementary particle with no electric charges.

Nike. A gas (KrF) laser facility at the Naval Research Laboratory.

Nova. A large glass laser facility at the Lawrence Livermore National Laboratory.

Nucleus. The positively charged core of an atom, composed of protons and neutrons. Protons and neutrons account for most of the atomic mass. The number of protons determines the chemical characteristics of an element.

Ohmic heating. Heating of the plasma by the plasma current due to the electrical resistance of the plasma.

Omega Upgrade. A large glass laser facility at the University of Rochester. Also called simply "Omega."

Orbit. As used here, the trajectory of an ion or electron moving in a magnetic field.

Particle-in-cell (PIC) computer program. A computer program that follows the motion of representative ions and electrons to generate detailed information about plasma behavior.

Passive safety systems. Techniques to shut down a fusion or fission reactor safely in the event of an accident without resort to mechanical devices that must be turned on in response to the accident.

Pellet. In ICF, the tiny target containing the DT fuel.

Petawatt. A very high power, short-pulse laser under development at the Lawrence Livermore National Laboratory. Also, a unit of power equal to 1,000 trillion watts.

Pinch. A plasma column carrying current that produces magnetic forces that constrict the column.

Plasma. A gaseous form of matter hot enough to become ionized, separating into positively charged ions and free electrons. Common examples are the glowing gases in fluorescent lights, and stars.

Plasma current. In a tokamak, current flowing toroidally around the plasma ring.

Pockels cell. An optical device used in the NIF beamline either to pass or to reflect the beam as needed in managing its progress through the amplifiers.

Poloidal field. In a tokamak, the component of magnetic field generated by the plasma current.

Potassium dihydrogen phosphate (KDP). Large crystals of this substance serve as efficient frequency convertors, turning near-infrared (1.05-micron) light into green light (in one step), and into 0.035-micron ultraviolet light (in two steps).

Pressure rule. As used here, an approximate condition used in designing a tokamak, relating the plasma pressure to the plasma current, the magnetic field, and the minor radius.

Princeton Large Torus (PLT). A neutral-beam-heated tokamak at the Princeton Plasma Physics Laboratory.

Propagating burn. The process by which fusion energy from an ignited hot spot spreads into and ignites neighboring cold fuel.

Proton. An elementary particle with positive charge; the nucleus of an ordinary hydrogen atom.

Pulse shaping. In ICF, the technique for increasing the driver power during the ablation phase so as to optimize the use of the driver energy to compress and heat the fuel.

Pumping. In glass lasers, the process of energizing the amplifiers by exciting metastable states of suitable elements, such as neodymium, incorporated in the amplifier glass.

Rayleigh-Taylor instability. A hydrodynamic instability occurring during the ICF pellet implosion process.

Rem. A unit of dose for any type of radiation, adjusted to be equivalent to depositing one-hundredth of a joule of x-ray energy per kilogram of body weight.

Resonance absorption. A laser-driven plasma instability in which light energy is absorbed as plasma waves.

Runaway electrons. In a tokamak, electrons accelerated to very high energies in certain circumstances.

Shallow burial. Disposal of radioactive waste that meets standards not requiring deep burial in underground depositories.

Spherical tokamak. A tokamak with roughly equal major and minor radii.

Spheromak. A toroidal magnetic plasma device in which the plasma itself creates closed magnetic flux surfaces like those of a tokamak.

Stellarator. A toroidal magnetic plasma device not requiring a plasma current.

Stimulated Brillouin scattering (SBS). One of several "parametric" plasma instabilities caused by the passage of intense light beams through a plasma.

Stimulated emission. In lasers, the process whereby a light beam causes excited atoms to emit light coherently, thereby amplifying the beam.

Stimulated Raman scattering (SRS). A "parametric" plasma instability, often generating suprathermal electrons.

Streak camera. A technique for recording measurements made over a period of time, adapted by ICF scientists to measure extremely rapid ablative implosion processes.

Superconducting magnets. Electromagnets made of materials that conduct electricity with negligible resistance.

Suprathermal electrons. Electrons accelerated to energies in excess of the average energy associated with the temperature.

Target. *See* Pellet.

Target chamber. The large chamber that contains the energy released by exploding pellets in an ICF reaction (see fig. 5).

Tesla. A unit of magnetic field strength equal to ten thousand gauss.

Thomson scattering. A technique for measuring the plasma electron temperature and density using lasers.

Tokamak. A type of magnetic confinement device; the main approach to magnetic fusion worldwide (see fig. 1).

Tokamak Fusion Test Reactor (TFTR). America's largest tokamak, at the Princeton Plasma Physics Laboratory.

Tokamak 3 (T-3). An early tokamak, at the Kurchatov Institute, Moscow.

Toroidal field. In a tokamak, the main magnetic field component, created by large magnetic coils wrapped around the vacuum vessel that contains the plasma.

Trigger beam. As used here, a weak light beam, generated by the master oscillator, that initiates a laser pulse.

Tritium. A radioactive isotope of hydrogen; one of the fuels used in a fusion reactor.

Turbulence. In plasmas, the presence of large-amplitude plasma waves that cause accelerated transport of heat and other effects important in designing fusion reactors.

2XIIB. A neutral-beam-heated magnetic mirror experiment at the Lawrence Livermore National Laboratory.

Vlasov equation. The basic theory of collective phenomena in plasmas.

Volt. A common unit of electric potential difference.

Watt. A unit of power equal to one joule per second.

Wavelength. A parameter characterizing light of different colors; the distance between crests, or peaks, in electric field strength in a propagating light wave.

Bibliography

Publications for a General Audience

Bromberg, Joan L. *Fusion.* Cambridge: MIT Press, 1982.

Craxton, R. Stephen, Robert L. McCrory, and John M. Soures. "Progress in Laser Fusion." *Scientific American,* August 1986, 68–79.

Fowler, T. K., and Richard F. Post. "Progress toward Fusion Power." *Scientific American,* December 1966, 21–31.

Furth, Harold P. "Fusion." *Scientific American,* September 1995, 174–76.

Heppenheimer, T. A. *The Man-Made Sun.* Boston: Little, Brown, 1984.

Herman, Robin. *Fusion: The Search for Endless Energy.* Cambridge: Cambridge University Press, 1990.

On-Line Information Sources

General Atomics, DIII-D Fusion Home Page, at http://FusionEd.gat.com/.

Lawrence Livermore National Laboratory, Inertial Confinement Fusion Research, X Division, at http://www-phys.llnl.gov/X_Div/.

Princeton Plasma Physics Laboratory, Fusion Energy Education Web Site, at http://FusEdWeb.pppl.gov/.

University of California, Berkeley, Department of Nuclear Engineering, Nuclear Fusion Section, at http://www.nuc.berkeley.edu/fusion/fusion.html.

U.S. Department of Energy, Fusion Energy Sciences Program, at http://wwwofe.er.doe.gov/.

Scientific Reviews of Fusion Technologies

Lindl, John. "Development of the Indirect-Drive Approach to Inertial Confinement Fusion and the Target Physics Basis for Ignition and Gain." *Physics of Plasmas* 2:3933–4024 (1995).

Sheffield, John. "The Physics of Magnetic Fusion Reactors." *Reviews of Modern Physics* 66:1015–1103 (1994).

Texts and Monographs

Hogan, W. J., ed. *Energy from Inertial Fusion.* Vienna: International Atomic Energy Agency, 1995.

Krall, Nicholas A., and Alvin W. Trivelpiece. *Principles of Plasma Physics.* New York: McGraw-Hill, 1973.

Simon, Albert. *An Introduction to Thermonuclear Research.* London: Pergamon Press, 1959.

Spitzer, Lyman, Jr. *Physics of Fully Ionized Gases.* New York: Interscience, 1956.

Stacey, Weston M., Jr. *Fusion: An Introduction to the Physics and Technology of Magnetic Confinement Fusion.* New York: Wiley-Interscience, 1981.

Teller, Edward, ed. *Fusion.* Vol. 1, pts. A, B. New York: Academic Press, 1981.

White, R. P. *Theory of Tokamak Plasmas.* Amsterdam: North-Holland, 1989.

Special Topics

Part I. Prometheus Unbound

EARLY FUSION

Atkinson, R. d'E., and F. G. Houtermans, "Zur frage der Aufbaumöglichkeit der Elemente in Sternen." *Zeitschrift für Physik* 54:656 (1929).

THE LAWSON CRITERION

Lawson, J. D. "Some Criteria for a Power Producing Thermonuclear Reactor." *Proceedings of the Physical Society of London,* sect. B, 70:6 (1957).

THE CURRENT RULE

Kruskal, M. D., and M. Schwarzchild. "Some Instabilities of a Completely Ionized Plasma." *Proceedings of the Royal Society of London,* ser. A, 223:348 (1954).

THE PRESSURE RULE

Troyon, F., R. Gruber, H. Saurenmann, S. Semenzato, and R. Succi. "MHD-Limits to Plasma Confinement." *Plasma Physics and Controlled Fusion* 26:209 (1984).

THE TOKAMAK T-3 "BREAKTHROUGH"

Artsimovich, L. A., et al. "Experiments in Tokamak Devices." In *Proceedings of the Third International Conference on Plasma Physics and Controlled Nuclear Fusion Research, Novosibirsk, 1968,* vol. I, p. 157. Vienna: International Atomic Energy Agency, 1969.

ENERGY CONFINEMENT TIME

Groebner, R. J. "An Emerging Understanding of H-Mode Discharges in Tokamaks." *Physics of Fluids B* 5:2343 (1993).

Kaye, S. M., and R. J. Goldston. "Global Energy Confinement Scaling for Neutral-Beam-Heated Tokamaks." *Nuclear Fusion* 25:65 (1985).

HIGH TEMPERATURES ACHIEVED

Eubank, H., and PLT Team. "PLT Neutral Beam Heating Results." In *Proceedings of the Seventh International Conference on Plasma Physics and Controlled Nuclear Fusion Research, Innsbruck, 1978,* vol. 1, p. 167. Vienna: International Atomic Energy Agency, 1979.

DEMONSTRATIONS OF FUSION POWER

Rebut, P. H., A. Gibson, M. Huguet, J. M. Adams, et al. "Fusion Energy Production from a Deuterium-Tritium Plasma in the JET Tokamak." *Nuclear Fusion* 32:187 (1992).

Strachen, J. D., and TFTR Team. "Fusion Power Production from TFTR Plasmas Fueled with Deuterium and Tritium." *Physical Review Letters* 72:3526 (1994).

Part II. Creating Fusion Science

MHD THEORY

Freidberg, J. P. "Ideal Magnetohydrodynamic Theory of Magnetic Fusion Systems." *Reviews of Modern Physics* 54:801 (1982).

Grad, H., and H. Rubin. "Hydromagnetic Equilibria and Force Free Fields." In *Proceedings of the UN International Conference on the Peaceful Uses of Atomic Energy,* vol. 31, p. 190. Geneva, 1958.

Shafranov, V. D. *Soviet Physics JETP* 6:545 (1958).

THE ENERGY PRINCIPLE

Bernstein, I. B., E. A. Frieman, M. D. Kruskal, and R. M. Kulsrud. "An Energy Principle for Hydromagnetic Stability Problems." *Proceedings of the Royal Society of London,* ser. A, 244:17 (1958).

FREE ENERGY

Fowler, T. K. "Bounds on Plasma Fluctuations and Anomalous Diffusion." *Physics of Fluids* 8:459 (1965).

———. "Thermodynamics of Unstable Plasmas." In *Advances in Plasma Physics,* edited by A. Simon and W. B. Thompson, vol. 1, pp. 201–25. New York: Interscience, 1968.

TRANSPORT AND TURBULENCE

Kadomtsev B. B., and O. P. Pogutse. "Trapped Particles in Toroidal Magnetic Systems." *Nuclear Fusion* 11:67 (1971).

Rechester, A. B., and M. N. Rosenbluth. "Electron Heat Transport in a Tokamak with Destroyed Magnetic Surfaces." *Physical Review Letters* 40:38 (1978).

Rosenbluth, M. N., R. D. Hazeltine, and F. L. Hinton. "Plasma Transport in Toroidal Confinement Systems." *Physics of Fluids* 15:116 (1972).

Shaing, K. C. "Neoclassical Quasilinear Transport Theory of Fluctuations in Toroidal Plasmas." *Physics of Fluids* 31:2249 (1988).

Tang, W. M. "Microinstability Theory in Tokamaks." *Nuclear Fusion* 18:1089 (1978).

PLASMA SELF-ORGANIZATION

Taylor, J. B. "Relaxation and Magnetic Reconnection in Plasmas." *Reviews of Modern Physics* 58:741 (1986).

———. "Relaxation of Toroidal Plasma and Generation of Reverse Magnetic Fields." *Physical Review Letters* 33:1139 (1974).

BOOTSTRAP CURRENT

Bickerton, R. J., J. W. Connor, and J. B. Taylor. "Diffusion Driven Plasma Currents and Bootstrap Tokamak." *Nature Physical Science* 229:110 (1971).

Nunan, W. J., and J. M. Dawson. "Computer Simulation of Transport Driven Current in Tokamaks." *Physical Review Letters* 73:1628 (1994).

Weening, R. H., and A. H. Boozer. "Completely Bootstrapped Tokamak." *Physics of Fluids B* 4:159 (1992).

MEASUREMENTS

Liewer, Paulett C. "Measurements of Microturbulence in Tokamaks and Comparisons with Theories of Turbulence and Anomalous Transport." *Nuclear Fusion* 25:543 (1985).

PLASMA TECHNOLOGY AND REACTORS

Conn, Robert W. "Magnetic Fusion Reactors." Chapter 14 in Teller, *Fusion.* (Teller, *Fusion,* is listed under "Texts and Monographs.")

Kunkel, W. B. "Neutral-Beam Injection." Chapter 12 in Teller, *Fusion.*

Moir, R. W. "The Fusion-Fission Fuel Factory." Chapter 15 in Teller, *Fusion.*

Porkolab, Miklos. "Radio-Frequency Heating of Magnetically Confined Plasma." Chapter 13 in Teller, *Fusion.*

Part III. Another Way: Inertial Fusion

LASERS

Maiman, T. H. "Stimulated Optical Radiation in Ruby." *Nature* 187:493 (1960).

LASER HEATING OF PLASMAS

Basov, N. G., and O. N. Krohkin. "The Conditions of Plasma Heating by the Optical Generator Method." In *Proceedings of the Third International Conference on Quantum Electronics, Paris, 1963*, pp. 1373–77. New York: Columbia University Press, 1964.

Dawson, J. M. "On Production of Plasma by Giant Pulse Lasers." *Physics of Fluids* 7:981 (1964).

EARLY INERTIAL CONFINEMENT FUSION

Bodner, S. E. "Critical Elements of High Gain Laser Fusion." *Journal of Fusion Energy* 1:221 (1981).

Brueckner, K. A., and S. Jorna. "Laser-Driven Fusion." *Reviews of Modern Physics* 46:325 (1974).

Kidder, R. E. "Energy Gain of Laser-Compressed Pellets: A Simple Model Calculation." *Nuclear Fusion* 16:405 (1976).

Meyer-ter-Vehn, J. "On Energy Gain of Fusion Targets: The Model of Kidder and Bodner Improved." *Nuclear Fusion* 22:561 (1982).

Nuckolls, John, Lowell Wood, Albert Thiessen, and George Zimmerman. "Laser Compression of Matter to Super-High Densities: Thermonuclear (CTR) Applications." *Nature* 239:139 (1972).

COMPUTATION

McCrory, R., and C. Verdon. "ICF Computer Simulations." In *Computer Applications in Plasma Science and Engineering*, edited by A. Drobat, pp. 291–325. Berlin: Springer-Verlag, 1991.

LASER-PLASMA INTERACTIONS

Kruer, William L. "Intense Laser Plasma Interactions: From Janus to Nova." *Physics of Fluids B* 3:2356 (1991).

LASER TECHNOLOGY

Bodner, S. E., R. Lehmberg, S. Obenschain, C. Pawley, M. Pronko, J. Sethian, A. Deniz, J. Hardgrove, T. Lehecka, O. Barr, I. Begio, S. Ozuchlewski, D. Hanson, N. Kurnit, W. Leland, J. McCleod, E. Rose, and G. York. "Krypton-Fluoride Laser Fusion Development in the USA." In *Plasma Physics and Controlled Nuclear Fusion Research, 1992*, vol. 3, p. 51. Vienna: International Atomic Energy Agency, 1993.

Emmett, J. L., W. F. Krupke, and J. B. Trenholme. "Future Development of High-Power Solid-State Laser Systems." *Soviet Journal of Quantum Electronics* 13:1 (1983).

Krupke, W. F. "Solid State Driver for an ICF Reactor." *Fusion Technology* 15:377 (1989).

Seka, W., S. D. Jacobs, J. E. Rizzo, R. Boni, and R. S. Craxton. "Demonstration of High Efficiency Third Harmonic Conversion of High Power Nd-Glass Laser Radiation." *Optics Communications* 34:469 (1980).

Soures, J. M., R. McCrory, T. Boehly, R. Craxton, S. Jacobs, J. Kelly, T. Kessler, J. Knauer, R. Knemens, S. Kumpan, S. Letzring, W. Seka, R. Short, M. Skeldon, S. Skupsky, and C. Verdon. "Omega Upgrade Laser for Direct-Drive Target Experiments." *Laser and Particle Beams* 11:317 (1991).

Van Wonterghem, B. M., J. R. Murray, D. R. Speck, and J. H. Campbell. "Performance of the NIF Prototype Beamlet." *Fusion Technology* 26:702 (1994).

ION DRIVERS

Reiser, M., T. Godlove, and R. Bangerter, eds. *Heavy Ion Inertial Fusion.* Conference Proceedings 152. New York: American Institute of Physics, 1986.

Vandevender, J. P., and D. L. Cook. "Inertial Confinement Fusion with Light Ions." *Science* 232:831 (1986).

THE LIQUID WALL BLANKET CONCEPT

Christofilos, N. C. "Design of a High Power-Density Astron Reactor." *Journal of Fusion Energy* 8:97 (1989). Report on work performed in 1970, published posthumously.

ICF REACTORS

Hogan, W., R. Bangerter, and G. Kulcinski. "Energy from Inertial Fusion." *Physics Today* 45:42 (1992).

Part IV. Fusion and Politics

FUSION POLICY REVIEWS

Holdren, John P., chair. *Fusion Review Panel Report,* by the President's Committee of Advisors on Science and Technology (PCAST). July 1995.

Stever, H. Guyford, chair. *Fusion Policy Advisory Committee Final Report.* DOE/S-0881. Washington, D.C.: U.S. Department of Energy, 1990.

THE DEMAND FOR ELECTRICITY

Rossin, A. David, and T. K. Fowler, eds. *Conversations about Electricity and the Future.* 1991. Available through American Nuclear Society, La Grange, Ill.

FUSION AND THE ENVIRONMENT

Dawson, J. M. "Advanced Fusion Reactors." Chapter 16 in E. Teller, *Fusion.* (Teller, *Fusion*, is listed under "Texts and Monographs.")

Holdren, J. P., chair, D. H. Berwald, R. J. Budnitz, J. G. Crocker, J. G. Delene, R. D. Endicott, M. S. Kazimi, R. A. Krakowski, B. G. Logan, and K. R. Schultz. "Report of the Senior Committee on Environmental, Safety, and Economic Aspects of Magnetic Fusion Energy." UCRL-53766. Lawrence Livermore National Laboratory report, September 25, 1989. (See also John P. Holdren et al., "Exploring the Competitive Potential of Magnetic Fusion Energy: The Interaction of Economics with Safety and Environmental Characteristics," *Fusion Technology* 13:7 [1988].)

ALTERNATIVE CONCEPTS

Baker, D. A., and W. E. Quinn. "The Reversed-Field Pinch." Chapter 7 in Teller, *Fusion.* (Teller, *Fusion*, is listed under "Texts and Monographs.")

Dandl, R. A., and G. E. Guest. "The ELMO Bumpy Torus." Chapter 11 in Teller, *Fusion.*

Fowler, T. K. "Mirror Theory." Chapter 5 in Teller, *Fusion.*

Post, R. F. "Experimental Base of Mirror-Confinement Physics." Chapter 6 in Teller, *Fusion.*

Quinn, W. E., and R. E. Siemon. "Linear Magnetic Fusion Systems." Chapter 8 in Teller, *Fusion.*

Ribe, F. L. "The High-Beta Stellarator." Chapter 9 in Teller, *Fusion.*

Ribe, F. L., and A. R. Sherwood. "Fast-Liner Compression Fusion Systems." Chapter 10 in Teller, *Fusion.*

Shoet, J. L. "Stellarators." Chapter 4 in Teller, *Fusion.*

Teller, Edward, Alexander J. Glass, T. Kenneth Fowler, Akira Hasegawa, and John F. Santarius. "Space Propulsion by Fusion in a Magnetic Dipole." *Fusion Technology* 22:82–97 (1992).

PREDICTABLE BREAKTHROUGHS

Fowler, T. K., J. S. Hardwick, and T. R. Jarboe. "On the Possibility of Ohmic Ignition in a Spheromak." *Comments on Plasma Physics and Controlled Fusion* 16:91 (1994).

Peng, Y.-K., and D. J. Strickler. "Features of Spherical Torus Plasmas." *Nuclear Fusion* 26:769 (1986).

Tabak, Max, James Hammer, Michael E. Glinsky, W. L. Kruer, Scott C. Wilks, John Woodworth, E. Michael Campbell, Michael D. Perry, and Rodney J. Mason. "Ignition and High Gain with Ultrapowerful Lasers." *Physics of Plasmas* 1:1626 (1994).

Index

United Technologies, 131
universal mode, 85
uranium, 4–5, 187, 194
U.S. Department of. *See other part of name*
U.S. Office of. *See other part of name*

vacuum vessel: ICF, 134; tokamak, 18
Vanderbilt University, 5
Vandevender, Pace, 153
Velikhov, Evgenii, 114–16, 194
Versailles, France, 114
Vienna, Austria, 116
Vlasov, A. A., 63
Vlasov equation, theory, 63–71, 82

Washington, University of, 206
Weening, Richard, 91

Weinberg, Alvin, 72, 85
Wisconsin, University of, 195
Wood, Lowell, 131
work, as free energy, 51, 77

xenon flashlamp, 159
x-rays, 21–22; in ICF, 140–41, 146, 155–56

yield, ICF, 135, 221
Yingzhong, Lu, 185
yin-yang magnet, 107–8, 177–78
Yoshikawa, Masagi, 95–96

Zabowski, Ron, 130
Zeta, 25, 48, 73, 87
Zimmerman, George, 132, 153

Library of Congress Cataloging-in-Publication Data

Fowler, T. Kenneth.
  The fusion quest / T. Kenneth Fowler.
       p.   cm.
  Includes bibliographical references and index.
  ISBN 0-8018-5456-3
       1. Controlled fusion.   2. Nuclear fusion.   3. Fusion reactors.
I. Title
  QC791.73.F69   1997
  621.48'4 — dc20   96-34254